Damiano Anselmi

RENORMALIZATION

Contents

Preface

Most high-energy physics, including the standard model of particle physics, is formulated by means of perturbative quantum field theory.

When the perturbative expansion is performed in a naïve way, it generates "divergences", that is to say, quantities that appear to be "infinite", instead of being small. Typically, they are due to diverging improper integrals. The presence of divergences suggests that it should be possible to define the perturbative expansion in a smarter way.

With the help of a cutoff, the divergences can become temporarily finite. Then they can be classified and moved around. Clearly, if a divergence disappears by changing the parametrization of the theory, it is not a true divergence, but just a blunder due to an unfortunate choice of variables. If there exists a reparametrization that makes all the divergences disappear, then the theory is actually convergent.

The divergences can be relocated by performing all sorts of operations that in normal circumstances leave the physics unchanged, such as changes of field variables, as well as redefinitions of parameters, in particular the coupling constants. Renormalization is the reparametrization that moves the divergences "to the right places", assuming that such places actually exist. In simple theories, the fields and the couplings just get multiplied by constants, whence the name re-*normalization*. In more complicated situations, the redefinitions can even be nonpolynomial.

Once the theory is renormalized, the cutoff can be safely removed, and the physical quantities become meaningful. Renormalization solves the problem of divergences, and allows us to define the correct perturbative expansion.

Under very general assumptions, it is *always* possible to absorb the divergences into reparametrizations. However, the price can be considerably high: the introduction of new physical parameters, which can in principle be infinitely many. If the divergences can be canceled by keeping the number of physical parameters finite, the theory is called renormalizable. Otherwise, it is called nonrenormalizable. The renormalizable theories acquire a very special status among all theories.

Renormalizability provides a criterion to select the physical theories among the huge set of theories that we can formulate mathematically. The selection is actually welcome, because it partially compensates for the difficulties we face when we explore the microscopic world, compared with the relative ease of exploring the macroscopic world. The class of renormalizable theories that we study in this book contains the standard model of particle physics in flat space. Therefore, it allows us to explain three interactions of nature out of four. No physical theory in more than four spacetime dimensions belongs to the same class, which makes renormalization a good candidate to explain why we live in four dimensions. However, quantum gravity does not belong to that class either. It can be included into a larger class of renormalizable theories, as long we accept that it is to some extent exceptional.

Inserting a parameter (the cutoff) to remove it later is a mathematical trick like many. In some sense, it is just a "technicality", and most of renormalization appears to be a rather technical issue. However, technicalities like this may have extremely important and unforeseen consequences, and considerably affect the physical predictions of the theory. Examples are offered by the renormalization-group flow, the particle widths and the anomalies: scale invariant theories can become scale dependent, coupling "constants" can become energy dependent, strong interactions can become weak, eternal particles can decay. The reason why the reparametrizations used to eliminate the divergences do not leave the physics completely unchanged is precisely that they are divergent.

Ironically, the "divergences" are the best known quantities of quantum field theory, to the extent that certain physical amplitudes can be calculated exactly to all orders because of their intimate relation with the divergences. At present, perturbative quantum field theory is the most successful theoretical achievement of elementary particle physics. Some of its aspects are so deep that most physicists need years and years to capture their true meanings. In some sense, the conceptual gap between quantum field theory and quantum mechanics can be compared to the one between quantum mechanics and classical mechanics. Several physicists have been puzzled by the uncertainty principle, and have never accepted that it could be part of the ultimate description of nature. Nowadays, some physicists still view the divergences as "pathologies", and think that "renormalization is a way to hide what *we* do not understand under the carpet". More probably, *they* do not understand what they are talking about.

Removing divergences is just a more sophisticated way to define improper integrals. Following Riemann, we can insert a cutoff, calculate the integral for finite values of the cutoff, and remove the cutoff at the end. If the procedure works, the integral is called convergent. If the procedure does not work, the integral is called divergent. Different prescriptions may lead to different results. For example, it is well known that the Riemann and the Lebesgue integrals are not equivalent.

Quantum field theory requires just one step more. There, we do not have one integral at a time, but a whole *theory*, which can be viewed as a collection of integrals, related to one another. We insert a cutoff Λ, calculate the integrals for finite values of Λ, and classify their divergent parts and their convergent parts. Instead of attempting to remove the cutoff in each integral separately, which in most cases makes no sense, we take advantage of the freedom we have for working with a collection of integrals. We make a variety of operations that *normally* do not change the physics, and inquire whether it is possible to relocate the divergent parts in order to make every integral meaningful *after* the relocation. If the procedure is successful, the theory is actually convergent. Otherwise, it is divergent.

In the end, we discover that the operations we make affect some physical predictions, and give results that crucially differ from what we expected at the beginning. However, we also discover that such changes are not only acceptable, but confirmed by experiments.

Ultimately, renormalization is one of the concepts we understand better, at present, in high-energy theoretical physics. Probably, all the future developments of high-energy physics will emerge more or less directly from it. At the same time, there is no doubt that quantum field theory is still formulated in a rather primitive way. A complete reformulation is desirable. One purpose of this book is to collect the basic knowledge about renormalization and stimulate people to start from that to upgrade the formulation of quantum field theory as much as it takes.

We are aware that in the past decades several approaches alternative to quantum field theory have been proposed, but we remain skeptical about their claimed virtues. Although they are often promoted as "beyond quantum field theory", the artificial enthusiasm that surrounded them for too long clashes with the poverty of their achievements. For example, there is little doubt that, conceptually speaking, string theory is a huge step backwards with respect to quantum field theory. We can only wish good luck to those who still do not see that all the alternatives to quantum field theory are doomed to sink into anonymity.

The book focuses on the fundamental aspects of renormalization. The main goals are to construct perturbative quantum field theory, study the consequences of renormalization, and show that the perturbative formulation of a wide class of theories is consistent to all orders. Most issues are treated using modern techniques, prioritizing the most economical and powerful tools. On the contrary, not much effort is devoted to explain how such a successful theoretical framework emerged historically. Some aspects of quantum field theory are very involved, and those who study the matter for the first time can greatly benefit from the rational, non-historical approach outlined here.

This book is not meant to replace the existing books on quantum field theory. Since its main focus is renormalization, several basic notions are just taken for granted. Quantum field theory is formulated by means of the functional integral, and the dimensional regularization technique, in the Euclidean framework. Algebraic aspects are covered to the extent that is necessary to treat renormalization.

Yet, the most popular textbooks fly over many important issues and details, which several readers, presumably, would like to be better explained to them. At least, they would appreciate being warned about crucial aspects that should not pass over in silence, and instead deserve to be adequately pointed out, to better assess the matter into the big picture. The ambition of this book is to fill the gaps. We could roughly describe it as "everything you always wanted to know about renormalization (and quantum field theory, to some extent), but nobody told you". In the same spirit, a nontrivial effort is spent to lay out even the most common subjects in ways that are different, and sometimes substantially different, from the popular ones. Finally, a number of exercises, with solutions, are distributed along the book, to help the layman familiarize with the most important tools of renormalization.

Chapter 1

Functional integral

The *functional integral* is an integral over a space of functions. It is one of the basic tools that are used to formulate the perturbative expansion of quantum field theory. It also provides an alternative formulation of quantum mechanics, which is equivalent to the Schrödinger and the Heisenberg ones.

The functional integral is defined as a limit of an ordinary multiple integral, when the number of integrated variables tends to infinity. Imagine that spacetime is discretized, with elementary cubic cells of size ℓ, and put into a box of finite size $L = N\ell$. The discretized pattern is called "lattice" and the distance ℓ between two vertices of the lattice is called "lattice space". For the moment, we work at finite values of ℓ and N, but at a second stage we take the limits $\ell \to 0$ and $N \to \infty$. At finite ℓ and N, the set of spacetime points x_i is finite and the discretized version of a function $f(x)$ is a finite set of values $f_i = f(x_i)$, with $i = 1, 2, \ldots N$. The f_i are the variables over which we integrate.

Consider the finite-dimensional ordinary integral

$$c(\ell, N) \int \prod_{i=1}^{N} \mathrm{d}f_i \; \hat{G}(f_i),\tag{1.1}$$

where $c(\ell, N)$ is a normalization factor, which can depend on ℓ and N, and $\hat{G}(f_i)$ is the discretized version of a generic functional $G(f)$. When ℓ tends to zero and N tends to infinity, the number of integrated variables tends to infinity. Assume that there exists a normalization factor $c(\ell, N)$ such that the limits $\ell \to 0$, $N \to \infty$ exist. Then, the functional integral over the space of functions $f(x)$ is defined as

$$\int [\mathrm{d}f]\, G(f) = \lim_{\substack{\ell \to 0 \\ N \to \infty}} c(\ell, N) \int \prod_{i=1}^{N} \mathrm{d}f_i\ \hat{G}(f_i).$$

The simplest integrals we need are Gaussian. The basic Gaussian multiple integral reads

$$\int_{-\infty}^{+\infty} \prod_{i=1}^{N} \mathrm{d}x_i\ \exp\left(-\frac{1}{2}\sum_{i,j=1}^{N} x_i M_{ij} x_j\right) = \frac{(2\pi)^{N/2}}{\sqrt{\det M}}, \tag{1.2}$$

where M is a positive-definite symmetric matrix. Formula (1.2) can be proved by diagonalizing M with an orthogonal matrix $\mathcal{N}$. Write $M = \mathcal{N} D \mathcal{N}^t$, where $D = \mathrm{diag}(m_1, \cdots, m_n)$ and m_i are the eigenvalues of M. Perform the change of variables $x = \mathcal{N}y$, and recall that the integration measure is invariant, since $\det \mathcal{N} = 1$. Then, the integral becomes the product of the one-dimensional Gaussian integrals

$$\int_{-\infty}^{+\infty} \mathrm{d}y_i\ \exp\left(-\frac{1}{2}m_i y_i^2\right) = \sqrt{\frac{2\pi}{m_i}},$$

whence (1.2) follows. We also have the formula

$$Z(a) = \int_{-\infty}^{+\infty} \prod_{i=1}^{N} \mathrm{d}x_i\ \exp\left(-\frac{1}{2}\sum_{i,j=1}^{N} x_i M_{ij} x_j + \sum_{i=1}^{N} x_i a_i\right)$$

$$= \frac{(2\pi)^{N/2}}{\sqrt{\det M}} \exp\left(\frac{1}{2}\sum_{i,j=1}^{N} a_i M_{ij}^{-1} a_j\right), \tag{1.3}$$

which can be easily proved from (1.2) by means of the translation $x = y + M^{-1}a$.

We can define the correlation functions

$$\langle x_{i_1} \cdots x_{i_n} \rangle = \frac{1}{Z(0)} \int_{-\infty}^{+\infty} \prod_{i=1}^{N} \mathrm{d}x_i\ x_{i_1} \cdots x_{i_n} \exp\left(-\frac{1}{2}\sum_{i,j=1}^{N} x_i M_{ij} x_j\right)$$

$$= \frac{1}{Z(a)} \frac{\partial^n Z(a)}{\partial a_{i_1} \cdots \partial a_{i_n}}\bigg|_{a=0}. \tag{1.4}$$

For example, we find

$$\langle x_j x_k \rangle = \frac{1}{Z(a)} \frac{\partial^2 Z(a)}{\partial a_j \partial a_k}\bigg|_{a=0} = M_{jk}^{-1},$$

$$\langle x_j x_k x_m x_n \rangle = M_{jk}^{-1} M_{mn}^{-1} + M_{jm}^{-1} M_{kn}^{-1} + M_{jn}^{-1} M_{km}^{-1}. \tag{1.5}$$

Every correlation function that contains an odd number of insertions vanishes: $\langle x_{i_1} \cdots x_{i_{2n+1}} \rangle = 0 \ \forall n$. Instead, the correlation functions that contain even numbers of insertions are determined by a simple recursion relation, which reads

$$\langle x_{i_1} \cdots x_{i_{2n}} \rangle = \sum_{k=2}^{2n} M_{i_1 i_k}^{-1} \langle x_{i_2} \cdots \widehat{x_{i_k}} \cdots x_{i_{2n}} \rangle. \tag{1.6}$$

where the hat denotes a missing insertion. This formula is proved by noting that

$$\langle x_{i_1} \cdots x_{i_{2n}} \rangle = \frac{1}{n!} \frac{\partial^{2n}}{\partial a_{i_1} \cdots \partial a_{i_{2n}}} \left(\frac{1}{2} a^t M^{-1} a \right)^n$$

$$= \frac{1}{2^{n-1}(n-1)!} \frac{\partial^{2n-1}}{\partial a_{i_2} \cdots \partial a_{i_{2n}}} \left[(M^{-1}a)_{i_1} \left(a^t M^{-1} a \right)^{n-1} \right]$$

$$= \frac{1}{2^{n-1}(n-1)!} \sum_{k=2}^{2n} M_{i_1 i_k}^{-1} \frac{\partial^{2(n-1)}}{\partial a_{i_2} \cdots \widehat{\partial a_{i_k}} \cdots \partial a_{i_{2n}}} \left(a^t M^{-1} a \right)^{n-1}$$

$$= \sum_{k=2}^{2n} M_{i_1 i_k}^{-1} \langle x_{i_2} \cdots \widehat{x_{i_k}} \cdots x_{i_{2n}} \rangle.$$

In the third line the hat on ∂a_{i_k} denotes a missing derivative. The recurrence relation (1.6) gives

$$\langle x_{i_1} \cdots x_{i_{2n}} \rangle = \sum_P M_{P(i_1)P(i_2)}^{-1} \cdots M_{P(i_{2n-1})P(i_{2n})}^{-1}, \tag{1.7}$$

where the sum is over the inequivalent permutations P of $\{i_1, \cdots i_{2n}\}$. By this we mean that identical contributions are counted only once.

Our first goal is to define the $\ell \to 0$, $N \to \infty$ limits of the multiple integrals just met, and others of similar types, and use them to formulate quantum mechanics and perturbative quantum field theory. We begin with quantum mechanics.

1.1   Path integral

Consider a non relativistic particle of mass m, potential $V(q)$ and Lagrangian

$$\mathcal{L}(q, \dot{q}) = \frac{m}{2}\dot{q}^2 - V(q). \tag{1.8}$$

Assume that the particle is observed in the location q_{in} at a time t_{in}, and in q_{f} at a time t_{f}. Also assume that it is not observed during the time interval $t_{\text{in}} < t < t_{\text{f}}$. Quantum mechanics teaches us that it is meaningless to tell "where" the particle is while it is not observed, or even assume that it is somewhere. In particular, it does not make sense to say that the particle moves from q_{in} to q_{f} along a particular trajectory $q(t)$, such as the classical trajectory that extremizes the action

$$S(q_{\text{f}}, t_{\text{f}}; q_{\text{in}}, t_{\text{in}}) = \int_{t_{\text{in}}}^{t_{\text{f}}} dt \ \mathcal{L}(q(t), \dot{q}(t)). \tag{1.9}$$

A tentative way out is to imagine that it moves from q_{in} to q_{f} along all the *paths*

$$q(t), \qquad t_{\text{in}} \leqslant t \leqslant t_{\text{f}}, \qquad q(t_{\text{in}}) = q_{\text{in}}, \qquad q(t_{\text{f}}) = q_{\text{f}},$$

at the same time. Then, each path must contribute to the physical quantities, with a suitable (complex) weight. This means that we have to somehow integrate over the paths.

In some sense, our plan is to replace the principle of minimum action with a new principle, which must account for the quantum effects. The semiclassical approximation suggests that each path should be weighed by the factor

$$\exp\left(\frac{i}{\hbar}S(q_{\text{f}}, t_{\text{f}}; q_{\text{in}}, t_{\text{in}})\right). \tag{1.10}$$

Indeed, when we study the limit $\hbar \to 0$, the strongly oscillating exponent singles out the trajectory of minimum action as the only one that survives.

Inside the path integral, we may need to damp the oscillations of (1.10) for large values of the action. To this purpose, we can assume that the mass m has a small positive imaginary part.

As mentioned earlier, one possibility to define the integral over the paths is to discretize the problem, and study the limit of an ordinary multiple integral, when the number of integrated variables tends to infinity. We discretize the time interval $t_{\text{in}} \leqslant t \leqslant t_{\text{f}}$ by dividing it in N subintervals

$$t_{i-1} \leqslant t \leqslant t_i, \qquad t_i = t_{i-1} + \varepsilon, \qquad \varepsilon = \frac{t_{\text{f}} - t_{\text{in}}}{N},$$

$i = 1, ..., N$, with $t_0 = t_{\text{in}}$ and $t_N = t_{\text{f}}$. The path $q(t)$ is then replaced by the set of positions $q_i = q(t_i)$ at the times t_i.

The trajectory of the i-th subinterval can be assumed to be the one that extremizes the action. However, in many cases simpler subtrajectories provide equally good approximations. For example, we can take the straight lines

$$\bar{q}(t) = \frac{q_i - q_{i-1}}{\varepsilon}(t - t_{i-1}) + q_{i-1}. \tag{1.11}$$

This choice leads to a picture like

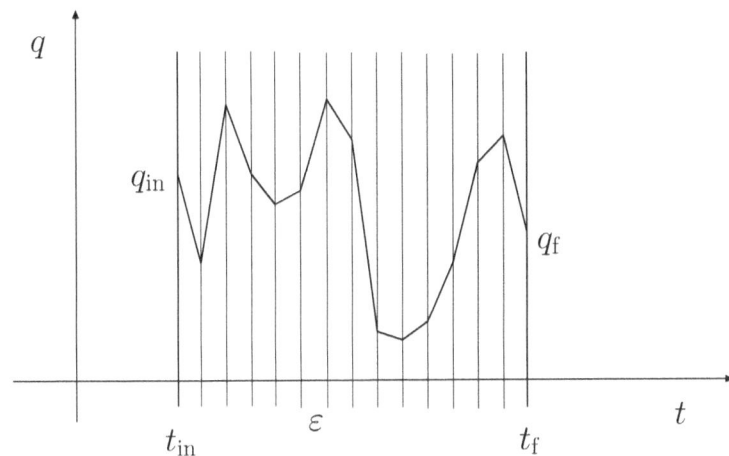

In the limit $\varepsilon \to 0$ the approximate path can tend to any function $q(t)$, including the ones that are not differentiable and not continuous. From the physical point of view, there is no reason why the unobservable trajectory $q(t)$ should be continuous and/or differentiable, so the path integral should indeed sum over all the functions $q(t)$.

In the i-th subinterval we have the constant velocity

$$\frac{q_i - q_{i-1}}{\varepsilon},$$

so the action (1.9) can be approximated by

$$\sum_{i=1}^{N} \bar{S}(q_i, t_i; q_{i-1}, t_{i-1}) = \sum_{i=1}^{N} \left\{ \frac{m(q_i - q_{i-1})^2}{2\varepsilon} - \varepsilon V(q_i) \right\} + \mathcal{O}(\varepsilon^{3/2}), \tag{1.12}$$

where the bar over S is there to remember that we have chosen the special subtrajectories (1.11). Below we prove that $|q_i - q_{i-1}| \sim \mathcal{O}(\varepsilon^{1/2})$, and that the corrections $\mathcal{O}(\varepsilon^{3/2})$ appearing in formula (1.12) can be neglected in the limit $\varepsilon \to 0$.

Inspired by (1.10), we weigh each infinitesimal portion of the trajectory by the factor

$$\frac{1}{A} \exp\left(\frac{i}{\hbar}\bar{S}(q_i, t_i; q_{i-1}, t_{i-1})\right),$$

where A is some normalization constant, to be determined. This means that, during a single time subinterval, the wave function $\psi(q, t)$ evolves into

$$\psi(q, t+\varepsilon) = \frac{1}{A} \int_{-\infty}^{+\infty} dq' \, \exp\left(\frac{i}{\hbar}\bar{S}(q, t+\varepsilon; q', t)\right) \psi(q', t). \qquad (1.13)$$

Consequently, during the finite interval $t_{\text{in}} \leqslant t \leqslant t_{\text{f}}$ the evolution of the wave function $\psi(q, t)$ is given by the formula

$$\psi(q, t) = \int_{-\infty}^{+\infty} dq' \, K(q, t; q', t')\psi(q', t'), \qquad (1.14)$$

where $K(q, t; q', t')$, called *kernel* of the time evolution, has the path-integral expression

$$K(q, t; q', t') = \lim_{N\to\infty} A^{-N} \int \prod_{i=1}^{N-1} dq_i \, e^{\frac{i}{\hbar}\sum_{i=1}^{N} \bar{S}(q_i, t_i; q_{i-1}, t_{i-1})} \qquad (1.15)$$

$$\equiv \int [dq] \exp\left(\frac{i}{\hbar}S(q)\right),$$

and now $t_0 = t'$, $t_N = t$, $q_0 = q'$, $q_N = q$, $\varepsilon = (t - t')/N$. The last line is the common short-hand notation used to denote the functional integral.

Observe that, in particular, we must have

$$K(q, t; q', t) = \delta(q - q'). \qquad (1.16)$$

Schrödinger equation

Now we prove that the time evolution encoded in the path-integral formulas (1.13) and (1.15) is equivalent to the one predicted by quantum mechanics. In particular, we show that the wave function (1.14), with the kernel defined by (1.15), satisfies the Schrödinger equation. We discretize time as explained above, and compare $\psi(q, t + \varepsilon)$ and $\psi(q', t)$ by means of (1.13). We have

$$\psi(q, t + \varepsilon) = \frac{1}{A} \int_{-\infty}^{+\infty} d\Delta \; e^{\frac{im\Delta^2}{2\hbar\varepsilon} - \frac{i\varepsilon}{\hbar} V(q) + \mathcal{O}(\varepsilon^{3/2})} \psi(q - \Delta, t), \qquad (1.17)$$

after a translation $q' = q - \Delta$. Recall that

$$\lim_{\varepsilon \to 0} \sqrt{\frac{m}{2\pi i \hbar \varepsilon}} e^{\frac{im\Delta^2}{2\hbar\varepsilon}} = \delta(\Delta).$$

This formula can be proved by assuming that the mass has a small positive imaginary part, as anticipated. Thus, the two sides of (1.17) match in the limit $\varepsilon \to 0$ if we take

$$\frac{1}{A} = \sqrt{\frac{m}{2\pi i \hbar \varepsilon}}.$$

Observe that this choice also ensures that (1.16) holds. We are still assuming that $|\Delta| \sim \varepsilon^{1/2}$, which allows us to neglect the Δ dependence contained inside $\mathcal{O}(\varepsilon^{3/2})$. This assumption is justified by the calculations that follow.

Expanding the integrand of (1.17) in powers of Δ, we obtain

$$\sqrt{\frac{m}{2\pi i \hbar \varepsilon}} \int_{-\infty}^{+\infty} d\Delta \; e^{\frac{im\Delta^2}{2\hbar\varepsilon}} \left(1 - \Delta \frac{\partial}{\partial q} + \frac{\Delta^2}{2} \frac{\partial^2}{\partial q^2} + \mathcal{O}(\Delta^3) \right.$$
$$\left. - \frac{i\varepsilon}{\hbar} V(q) + \mathcal{O}(\varepsilon^{3/2}) \right) \psi(q, t). \qquad (1.18)$$

Defining the integrals

$$I_n = \int_{-\infty}^{+\infty} d\Delta \; \Delta^n e^{\frac{im\Delta^2}{2\hbar\varepsilon}},$$

we find $I_{2k+1} = 0$ and

$$I_{2k} = -2i\varepsilon\hbar \frac{\partial I_{2k-2}}{\partial m}, \qquad I_0 = \sqrt{\frac{2\pi i \hbar \varepsilon}{m}},$$

which gives

$$I_{2k} = \Gamma\left(k + \frac{1}{2} \right) \left(\frac{2i\hbar\varepsilon}{m} \right)^{k+\frac{1}{2}}.$$

We find $I_{2k}/I_0 \sim \varepsilon^k$, which also proves $|\Delta| \sim \varepsilon^{1/2}$, as claimed before. Finally, rearranging (1.17), dividing by ε and taking the limit $\varepsilon \to 0$, we find the Schrödinger equation

$$i\hbar \frac{\partial \psi}{\partial t} = -\frac{\hbar^2}{2m} \frac{\partial^2 \psi}{\partial q^2} + V\psi.$$

The outcome is independent of the approximation we have used to expand $\bar{S}(q_i, t_i; q_{i-1}, t_{i-1})$. For example, we could have written $V(q_{i-1})$ in (1.12), instead of $V(q_i)$, or $(V(q_i) + V(q_{i-1}))/2$. The difference is always made of terms that are $\mathcal{O}(\varepsilon^{3/2})$ in the integrand of (1.18), which are negligible in the limit $\varepsilon \to 0$.

To conclude, we have proved that the path integral provides a formulation of quantum mechanics that is equivalent to the Schrödinger and the Heisenberg ones.

Free particle

We explicitly calculate the kernel in the case of the free particle. There,

$$\bar{S}(q_i, t_i; q_{i-1}, t_{i-1}) = \frac{m(q_i - q_{i-1})^2}{2\varepsilon},$$

so we have

$$K_{\text{free}}(q, t; q', t') = \lim_{N\to\infty} \left(\frac{m}{2\pi i \hbar\varepsilon}\right)^{N/2} \int \prod_{i=1}^{N-1} dq_i \; e^{\frac{im}{2\hbar\varepsilon}\sum_{i=1}^{N}(q_i-q_{i-1})^2}.$$

Changing variables to $\tilde{q}_i = q_i - q$, we can rewrite the integral as

$$e^{\frac{im}{2\hbar\varepsilon}(q-q')^2} \int \prod_{i=1}^{N-1} d\tilde{q}_i \; e^{\frac{im}{2\hbar\varepsilon}\left(\tilde{q}^t \tilde{M}\tilde{q} + 2\tilde{q}_1(q-q')\right)},$$

where

$$\tilde{M} = \begin{pmatrix} 2 & -1 & 0 & 0 & \cdots \\ -1 & 2 & -1 & \cdots & 0 \\ 0 & -1 & \cdots & -1 & 0 \\ 0 & \cdots & -1 & 2 & -1 \\ \cdots & 0 & 0 & -1 & 2 \end{pmatrix} \tag{1.19}$$

is an $(N-1) \times (N-1)$ matrix. Now the integral is of the Gaussian form (1.3) with

$$M = -\frac{im}{\hbar\varepsilon}\tilde{M}, \qquad a = \frac{im(q-q')}{\hbar\varepsilon}(0, \ldots 0, 1)$$

and $N \to N-1$. Again, we assume that the mass has a small positive imaginary part. We have

$$\det \tilde{M} = N, \qquad (\tilde{M}^{-1})_{N-1, N-1} = \frac{N-1}{N}. \tag{1.20}$$

The first formula can be proved recursively. Indeed, denoting the $I \times I$ matrix $\tilde{M}$ of (1.19) with $\tilde{M}_I$, we have

$$\det \tilde{M}_{N-1} = 2 \det \tilde{M}_{N-2} - \det \tilde{M}_{N-3}, \qquad \det \tilde{M}_1 = 2 \qquad \det \tilde{M}_2 = 3.$$

The second formula of (1.20) gives the last entry of the inverse matrix $\tilde{M}_{N-1}^{-1}$, and is just the determinant of the associated minor (which coincides with $\tilde{M}_{N-2}$), divided by the determinant of $\tilde{M}_{N-1}$.

Finally, using formula (1.3), with the appropriate substitutions, and recalling that $\varepsilon N = t - t'$, we find

$$K_{\text{free}}(q, t; q', t') = \sqrt{\frac{m}{2\pi i \hbar (t - t')}} \, e^{\frac{im(q-q')^2}{2\hbar(t-t')}},$$

which is the known result.

1.2   Free field theory

Given a classical field theory, described by the action $S(\varphi)$, we want to define the functional integral

$$\int [\mathrm{d}\varphi] \, \exp\left(\frac{i}{\hbar} S(\varphi)\right). \tag{1.21}$$

At present, we can do this only perturbatively, by expanding around the free-field limit.

From now on, we work in Euclidean space, where some complications are avoided. For simplicity, we also set $\hbar = 1$.

Free field theories are described by Gaussian functional integrals. We start from the scalar field in four dimensions. Its action in Euclidean space is

$$S(\varphi) = \frac{1}{2} \int \mathrm{d}^4 x \left((\partial_\mu \varphi)^2 + m^2 \varphi^2 \right). \tag{1.22}$$

We want to define the generating functional

$$Z(J) \equiv e^{W(J)} = \int [\mathrm{d}\varphi] \, \exp\left(-S(\varphi) + \int \varphi J\right) \tag{1.23}$$

where J are external sources, $\int \varphi J \equiv \int \mathrm{d}^4 x \, \varphi(x) J(x)$ and W is the logarithm of Z. First, we discretize the Euclidean space. Each coordinate x^μ is replaced by an index i_μ, and the field $\varphi(x)$ becomes $\varphi_{\{i_\mu\}}$. The discretized form of the action reads

$$S_{\text{discr}}(\varphi_{\{i_\mu\}}) = \frac{1}{2} \sum_{\{i_\mu\},\{j_\nu\}} \varphi_{\{i_\mu\}} M_{\{i_\mu\}\{j_\nu\}} \varphi_{\{j_\nu\}}, \qquad (1.24)$$

for some matrix $M_{\{i_\mu\}\{j_\nu\}}$ that we do not need to specify here. It suffices to recall that M is the discretized version of the kinetic operator $-\Box + m^2$:

$$M_{\{i_\mu\}\{j_\nu\}} = (-\Box + m^2)\big|_{\text{discr}}. \qquad (1.25)$$

The discretized version of the functional integral reads

$$Z(J_{\{i_\mu\}}) = \int \prod_{\{i_\mu\}} [\mathrm{d}\varphi_{\{i_\mu\}}] \, \exp\left(-S_{\text{discr}}(\varphi_{\{i_\mu\}}) + \sum_{\{i_\mu\}} \varphi_{\{i_\mu\}} J_{\{i_\mu\}}\right).$$

Using formula (1.3), we find

$$W(J_{\{i_\mu\}}) = \ln Z(J_{\{i_\mu\}}) = \frac{1}{2} \sum_{\{i_\mu\},\{j_\nu\}} J_{\{i_\mu\}} M^{-1}_{\{i_\mu\}\{j_\nu\}} J_{\{j_\nu\}} - \frac{1}{2} \ln \det M + C,$$

$$(1.26)$$

where C is a constant that includes the normalization factor $c(\ell, L)$ of formula (1.1).

To define the continuum limit, we basically need to define the inverse of M and its determinant. However, note that the determinant, as well as the constant C, appears only in Z and W, but not in the correlation functions (1.4). Therefore, we actually just need to define M^{-1}. This is not difficult, because the inverse of $-\Box + m^2$ is by definition the Green function $G_B(x, y)$, that is to say, the solution of the equation

$$(-\Box_x + m^2)G_B(x, y) = \delta^{(4)}(x - y). \qquad (1.27)$$

Normalizing the functional integral conveniently and using (1.3) and (1.26), we can write

$$Z(J) = e^{W(J)}, \qquad W(J) = \frac{1}{2} \int \mathrm{d}^4x \, J(x) G_B(x, y) J(y) \mathrm{d}^4y. \qquad (1.28)$$

We can define the J-dependent correlation functions

$$\langle \varphi(x_1) \cdots \varphi(x_n) \rangle_J = \frac{\int [\mathrm{d}\varphi] \varphi(x_1) \cdots \varphi(x_n) \exp\left(-S(\varphi) + \int \varphi J\right)}{\int [\mathrm{d}\varphi] \exp\left(-S(\varphi) + \int \varphi J\right)}$$

$$= \frac{1}{Z(J)} \frac{\delta^n Z(J)}{\delta J(x_1) \cdots \delta J(x_n)}, \qquad (1.29)$$

where the subscript J means that the sources are nonvanishing. The correlation functions at $J = 0$ are simply written as $\langle \varphi(x_1) \cdots \varphi(x_n) \rangle$. In particular, formulas (1.5) give

$$\langle \varphi(x)\varphi(y) \rangle = \frac{1}{Z(J)} \frac{\partial^2 Z(J)}{\partial J(x)\partial J(y)} \bigg|_{J=0} = G_B(x, y), \qquad (1.30)$$

$$\langle \varphi(x)\varphi(y)\varphi(z)\varphi(w) \rangle = G_B(x, y)G_B(z, w) + G_B(x, z)G_B(y, w)$$
$$+ G_B(x, w)G_B(y, z), \qquad (1.31)$$

and so on. We see that, in practice, the free theory contains just one piece of information, which is the Green function.

Formulas (1.6) and (1.7) can be generalized by following the same steps. We have

$$\langle \varphi(x_1) \cdots \varphi(x_{2n}) \rangle = \sum_{k=2}^{2n} G_B(x_1, x_k) \langle \varphi(x_2) \cdots \widehat{\varphi(x_k)} \cdots \varphi(x_{2n}) \rangle \qquad (1.32)$$

$$= \sum_P G_B(x_{P(1)}, x_{P(2)}) \cdots G_B(x_{P(2n-1)}, x_{P(2n)}), \qquad (1.33)$$

while the correlation functions that contain an odd number of insertions vanish.

Equation (1.33) is known as *Wick's theorem*. It says that i) the external points $x_1 \cdots x_{2n}$ must be connected pairwise in all the inequivalent ways, ii) each connection is a Green function and iii) each inequivalent set of connections is multiplied by the coefficient 1.

It is natural to express Wick's theorem graphically. A Green function is drawn as a double line connecting a pair of points. Then formula (1.32) reads

$$(1.34)$$

where the legs attached to the discs denote the insertions of the correlation functions.

The Euclidean Green functions can be computed by switching to momentum space. We define the Fourier transform as

$$\varphi(x) = \int \frac{\mathrm{d}^4 p}{(2\pi)^4} e^{ip \cdot x} \tilde{\varphi}(p). \tag{1.35}$$

Then we find

$$G_B(x, y) = \langle \varphi(x)\varphi(y) \rangle = \int \frac{\mathrm{d}^4 p}{(2\pi)^4} \frac{e^{ip \cdot (x-y)}}{p^2 + m^2} = \frac{m}{4\pi^2 |x - y|} K_1(m|x - y|),$$
$$\tag{1.36}$$

where K_1 denotes the modified Bessel function of the second kind.

This result is proved as follows. We must assume that x and y do not coincide, and use a trick to remove the oscillating behavior at infinity. Let us start from the massless limit. To calculate the integral at $m = 0$, we multiply the integrand by $e^{-\delta|p|}$, where $\delta > 0$, and take the limit $\delta \to 0$ at the end. Switching to spherical coordinates, we first integrate over $|p|$ and later over the angles. The basic steps are

$$G_B(x, 0)|_{m=0} = \lim_{\delta \to 0^+} \int \frac{\mathrm{d}^4 p}{(2\pi)^4} \frac{e^{ip \cdot x - \delta|p|}}{p^2} = \lim_{\delta \to 0^+} \frac{1}{4\pi^3} \int_0^\pi \frac{\mathrm{d}\theta \sin^2 \theta}{(\delta - i|x| \cos \theta)^2}$$
$$= \lim_{\delta \to 0^+} \frac{1}{4\pi^2 x^2} \left(1 - \frac{\delta}{\sqrt{\delta^2 + x^2}} \right) = \frac{1}{4\pi^2 x^2}. \tag{1.37}$$

To calculate the integral at $m \neq 0$, we make it convergent in a different way (at $x \neq 0$), that is to say by differentiating with respect to m. Then, after switching to spherical coordinates, we first integrate over the angles, and later over $|p|$. We find

$$\frac{\partial}{\partial m} G_B(x, 0) = - \int \frac{\mathrm{d}^4 p}{(2\pi)^4} \frac{2m e^{ip \cdot x}}{(p^2 + m^2)^2}$$
$$= -\frac{m}{2\pi^2 |x|} \int_0^\infty \frac{p^2 J_1(p|x|) \, \mathrm{d}p}{(p^2 + m^2)^2} = -\frac{m}{4\pi^2} K_0(m|x|),$$

where $J_1(x)$ is the Bessel function of the first kind. Integrating over m and requiring (1.37) at $m = 0$, we obtain (1.36).

The correlations functions can be mathematically interpreted as distributions. Then, the sources J should be viewed as test functions. Indeed, the

Green function $G_B(x,y)$, which appears to be singular at $x = y$, is actually regular as a distribution. To see this, it is sufficient to observe that when $G_B(x,y)$ acts on a test function $J(y)$, the behavior of the integral

$$\int_{|x-y|\sim 0} \mathrm{d}^4 y\, G_B(x,y) J(y) \sim \frac{1}{4\pi^2} \int \mathrm{d}^4 y\, \frac{J(y)}{(x-y)^2},$$

around $x \sim y$ shows no singularity. We have used the result (1.37), since the behavior of (1.36) at $x \sim y$ coincides with the behavior at $m \to 0$.

1.3 Perturbative expansion

An interacting theory can be defined by expanding perturbatively around its free-field limit. Although this sounds like a straightforward process, the perturbative expansion of quantum field theory underlies a huge conceptual advancement with respect to the notions we are accustomed to. To clarify this point, it is worth paying attention to what we do when we normally approximate. We have, say, difficult differential equations, which we need to solve for some physical applications. We know classes of exact solutions, but typically they do not cover enough cases of physical interest. We realize that some physical situations are only slightly different from those described by the exact solutions, so we work out other solutions by expanding perturbatively around the exact ones. What is important for the present discussion, is that we are talking about a well defined problem, described by difficult, but well defined, equations. *Then*, we approximate. We approximate something that does exist, something that exists *before* the approximation.

In quantum field theory, instead, we must really start from nothing, apart from the free-field limit. There are no equations, and no theory, before we make approximations. Thus, when we say that we perturbatively expand around the free field theory, we are actually lying: we are not expanding at all. The truth is that we are perturbatively *building* the interacting theory, piece by piece, out of the free field one. The endeavor we are going to undertake is a creative one, not just a deductive process. Therefore, if something goes wrong along the way, it will be no real surprise. To solve the problems that emerge, we have to be more and more creative. In particular, we have to build the mathematics that we need along the way, by ourselves. Not only, but every time we find a difficulty, and guess a possible solution, we must start over, implement the proposed solution from the very beginning, and rederive everything up to the point where we found the problem, check that the problem does disappear as expected, and ensure that no collateral difficulties emerge.

Another crucial point is that the perturbative expansion should be considered as a formal power series. In mathematics, a formal power series is a power series that is just viewed as a list of addends, disregarding completely whether the sum converges or not. Perturbative quantum field theory investigates the consistency of the perturbative expansion as a formal power series. It studies the properties of the addends (e.g., their consistency with gauge invariance, renormalization, unitarity, the cancellation of anomalies, etc.) and the relations among them. Proving that, for example, the Standard Model is consistent to all orders, as a perturbative quantum field theory, is already a nontrivial task. The great advantage of working with formal power series is that it allows us to freely exchange the sum with derivative operations, as well as integral operations. Only at the very end we will inquire whether the sum converges or not. Indeed, it is meaningless to demand that a power series be convergent before having shown that it obeys all the desirable physical and mathematical requirements as a formal powers series. The renormalization group, the particle widths and the anomalies of quantum field theory provide well-known cases where the power series are in the end convergent.

Having to build some of the mathematics anew is not surprising either. Normally, we take for granted that the mathematics we already know is good enough to formulate the laws of physics that apply to unexplored research domains. However, this assumption is too restrictive. More reasonably, our mathematics is necessarily limited, because it is a product of the interaction between us and the environment in which we are placed as human beings. When we explore energy domains that are very different from those we are accustomed to, we do not know whether the mathematics we have previously developed is good enough or not. For example, no difficulty like the "systematic creative approximation out of nowhere" has emerged, so far, in the studies of cosmic and astrophysical phenomena. On the other hand, the investigation of microscopic phenomena has triggered unforeseen challenges. In the case of quantum mechanics, we have been able to fill the gap, to some extent, by means of the "correspondence principle". The idea was that, although there are huge differences between the classical phenomena and the quantum ones, at least there is a sort of correspondence between the two. Clearly, we cannot expect to go on forever by relying on lucky correspondences, to the extent that quantum field theory forces us to abandon that idea. For example, the "classical" Lagrangian of quantum chromodynamics, which is the theory that describes what the strong interactions look like at high energies, has no correspondence with the classical phenomena. Exploring smaller and smaller distances, the problem becomes harder and harder, and we may be forced to give up every correspondence with what we know, and even renounce common sense and intuition, to develop a completely new mathematics by plunging into pure abstraction and technicalism. Quantum field theory, renormalization, the problems we find along the way and the partial solutions we manage to work out, give us hints of what the new mathematics will have to be.

That said, the only thing we can do at the moment is pretend there is nothing to worry about, and make a step forward along the process of "creative approximation".

Consider a theory of interacting scalar fields with action $S(\varphi) = S_0(\varphi) + S_I(\varphi)$, where $S_0(\varphi)$ is (1.22). For concreteness, we can take the φ^4 theory in four dimensions, which has

$$S(\varphi) = \int d^4x \left(\frac{1}{2}(\partial_\mu \varphi)^2 + \frac{m^2}{2}\varphi^2 + \frac{\lambda}{4!}\varphi^4 \right). \qquad (1.38)$$

Defining $Z(J)$ as in (1.23), we can write

$$\begin{aligned}
Z(J) &= \int [d\varphi] \, \exp\left(-S(\varphi) + \int J\varphi \right) \\
&= \int [d\varphi] \, \exp\left(-S_I(\varphi) \right) \exp\left(-S_0(\varphi) + \int J\varphi \right) \\
&= \sum_{n=0}^{\infty} \frac{(-1)^n}{n!} \int [d\varphi] \, S_I^n(\varphi) \exp\left(-S_0(\varphi) + \int J\varphi \right) \\
&= Z_0(J) \sum_{n=0}^{\infty} \frac{(-1)^n}{n!} \langle S_I^n(\varphi) \rangle_{0,J},
\end{aligned} \qquad (1.39)$$

where $Z_0(J)$ is given by (1.28). We use the subscript 0 to denote quantities at $\lambda = 0$. In particular, $\langle \cdots \rangle_{0,J}$ are the free-field correlation functions at nonvanishing sources. We have

$$\langle S_I^n(\varphi)\rangle_{0,J} = \left(\frac{\lambda}{4!}\right)^n \int \prod_{i=1}^n \mathrm{d}^4 x_i \, \langle \varphi^4(x_1) \cdots \varphi^4(x_n)\rangle_{0,J}.$$

Now, by (1.29) every φ-insertion can be expressed as a functional derivative with respect to J. Therefore,

$$\langle S_I^n(\varphi)\rangle_{0,J} = \left(\frac{\lambda}{4!}\right)^n \frac{1}{Z_0(J)} \int \prod_{i=1}^n \mathrm{d}^4 x_i \, \frac{\delta^{4n} Z_0(J)}{\delta J^4(x_1) \cdots \delta J^4(x_n)}.$$

Inserting this formula into (1.39), we get

$$Z(J) = \sum_{n=0}^{\infty} \frac{1}{n!} \left(-\frac{\lambda}{4!}\right)^n \prod_{i=1}^n \left(\int \mathrm{d}^4 x_i \frac{\delta^4}{\delta J^4(x_i)}\right) Z_0(J)$$

$$= \exp\left(-\frac{\lambda}{4!} \int \mathrm{d}^4 x \frac{\delta^4}{\delta J^4(x)}\right) Z_0(J).$$

More generally, we have

$$Z(J) = \mathrm{e}^{W(J)} = \exp\left(-S_I\left(\frac{\delta}{\delta J}\right)\right) Z_0(J). \tag{1.40}$$

The scalar field inside S_I is formally replaced by the functional derivative $\delta/\delta J$, which acts on the free-field generating functional $Z_0(J)$.

Formula (1.40) expresses the generating functional of the interacting theory as an infinite sum of terms, each of which involves just functional derivatives of the generating functional of the free field theory (which, as we know, contains only the Green function), and integrals over coordinates. Some functional derivatives are evaluated at the same point, which is called "vertex". Moreover, the Green functions connect pairs of points, as we see from the Wick theorem (1.34).

Formula (1.40) can be efficiently expressed diagrammatically. Diagrams are made of vertices and lines, and are drawn by following a simple set of rules, which we now derive.

Feynman rules

The correlation functions can be defined from the expansion of the generating functional $Z(J)$ in powers of J:

$$Z(J) = Z(0) \sum_{n=0}^{\infty} \frac{1}{n!} \int \left(\prod_{i=1}^n \mathrm{d}^4 x_i\right) \langle \varphi(x_1) \cdots \varphi(x_n)\rangle \, J(x_1) \cdots J(x_n).$$

For some practical purposes, it is also useful to define correlation functions with a different normalization. At $J \neq 0$ we define

$$\langle \varphi(x_1) \cdots \varphi(x_n) \rangle'_J = \frac{1}{Z_0(0)} \frac{\delta^n Z(J)}{\delta J(x_1) \cdots \delta J(x_n)}, \qquad (1.41)$$

while at $J = 0$ we write these objects as $\langle \varphi(x_1) \cdots \varphi(x_n) \rangle'$. In particular, we have

$$\langle \varphi(x_1) \cdots \varphi(x_n) \rangle = \frac{Z_0(0)}{Z(0)} \langle \varphi(x_1) \cdots \varphi(x_n) \rangle' = \frac{\langle \varphi(x_1) \cdots \varphi(x_n) \rangle'}{\langle 1 \rangle'}. \qquad (1.42)$$

Since we have normalized $Z_0(0)$ to 1, we could omit this factor. However, the formulas are more explicit if we keep it, which also emphasizes that $Z(0)$ is not equal to one.

Observe that $Z(J)$ can also be viewed as the generating functional $Z'(J)$ of the correlation functions (1.41):

$$Z'(J) = Z_0(0) \sum_{n=0}^{\infty} \frac{1}{n!} \int \left(\prod_{i=1}^{n} d^4 x_i \right) \langle \varphi(x_1) \cdots \varphi(x_n) \rangle' J(x_1) \cdots J(x_n) = Z(J). \qquad (1.43)$$

Consider a generic correlation function (1.41) of the φ^4 theory. Writing

$$\langle \varphi(x_1) \cdots \varphi(x_n) \rangle' = \frac{1}{Z_0(0)} \int [d\varphi] \varphi(x_1) \cdots \varphi(x_n) e^{-S_0(\varphi) - \frac{\lambda}{4!} \int d^4 x \, \varphi^4(x)}$$

$$= \sum_{k=0}^{\infty} \frac{(-\lambda)^k}{(4!)^k k!} \int \left(\prod_{j=1}^{k} d^4 y_j \right) \langle \varphi(x_1) \cdots \varphi(x_n) \prod_{j=1}^{k} \varphi^4(y_j) \rangle_0, \qquad (1.44)$$

we obtain a sum of contributions that are due to free-field correlation functions with $n+4k$ insertions. We call the points $x_1, \cdots, x_n$ "external" and the points $y_1, \cdots, y_k$ "internal". Each internal point carries four φ insertions. We refer to it as a vertex with four legs.

The free-field correlation functions of (1.44) can be worked out by means of Wick's theorem. Let us consider the graphical version (1.34) of that theorem. We see that each point is connected once to every other point. Moreover, each contribution is multiplied by the coefficient one. Thus, the interacting correlation function (1.44) is expressed as a sum of diagrams that are constructed by applying the following rules:

1) the diagrams have n external points $x_1, \cdots, x_n$ and an arbitrary number k of internal points $y_1, \cdots, y_k$; the latter are called vertices;

2) lines connect pairs of points; a line is called internal if it connects two internal points, otherwise it is called external;

3) the line that connects two points z and w is associated with the Green function $G_B(z, w)$;

4) four legs are attached to each internal point, one leg to each external point;

5) each diagram with k vertices is multiplied by the factor

$$\frac{1}{k!} \left(\frac{-\lambda}{4!} \right)^k ; \tag{1.45}$$

6) the positions y of the vertices are integrated with measure $\mathrm{d}^4 y$.

For example, consider the case $n = 2$, $k = 0, 1$. We have

$$\langle \varphi(x_1)\varphi(x_2) \rangle' = G_B(x_1, x_2) - \frac{\lambda}{4!} \int \mathrm{d}^4 y \, \langle \varphi(x_1)\varphi(x_2)\varphi^4(y) \rangle_0 + \mathcal{O}(\lambda^2)$$

$$= G_B(x_1, x_2) - \frac{\lambda}{2} \int \mathrm{d}^4 y \, G_B(x_1, y) G_B(y, y) G_B(y, x_2)$$

$$- \frac{\lambda}{8} G_B(x_1, x_2) \int \mathrm{d}^4 y \, G_B^2(y, y) + \mathcal{O}(\lambda^2),$$

which graphically reads

$$\tag{1.46}$$

plus $\mathcal{O}(\lambda^2)$.

Different contributions originated by the right-hand side of (1.34) can give the same diagram, that is to say, the same integral. For example, the second diagram on the right-hand side of (1.46) appears 12 times, which is why its coefficient is in the end $1/2$. Instead, the third diagram appears 3 times, so its coefficient is $1/8$.

We can collect the arrangements that give the same diagram into a single contribution, provided we multiply it by a suitable combinatorial factor. Then, the perturbative expansion is organized as a sum over inequivalent diagrams G, which are multiplied by (1.45) and an extra factor s_G, which counts how many contributions of Wick's theorem give the same G.

It is also convenient to switch to momentum space, where some further simplifications occur. For example, consider the last-but-one diagram of (1.46). We find

$$
\int d^4u \; G_B(x,u)G_B(u,u)G_B(u,y)
$$

$$
= \int d^4u \int \frac{d^4p}{(2\pi)^4} \frac{d^4k}{(2\pi)^4} \frac{d^4q}{(2\pi)^4} \frac{e^{ip(x-u)+iq(u-y)}}{(p^2+m^2)(k^2+m^2)(q^2+m^2)}. \quad (1.47)
$$

The u-integral can be evaluated immediately, and gives $(2\pi)^4\delta^{(4)}(p-q)$. Thus, (1.47) is the Fourier transform of

$$
(2\pi)^4\delta^{(4)}(p-q)\frac{1}{p^2+m^2}\left(\int \frac{d^4k}{(2\pi)^4}\frac{1}{k^2+m^2}\right)\frac{1}{q^2+m^2} \quad (1.48)
$$

on p and q. This formula illustrates some properties that are actually valid for all the graphs. First, we learn that it is much more convenient to work in momentum space, rather than in coordinate space. Indeed, (1.48) looks much simpler than the left-hand side of (1.47). Second, the theory is invariant under translations, so the total momentum is conserved. As a consequence, each correlation function is multiplied by a delta function like the one appearing in (1.48), which ensures that the momentum that enters the graph equals the momentum that exits from it, or, equivalently, that the total momentum that enters the graph vanishes. We do not need to write this delta function explicitly every time, so from now on we simply omit it. Third, the factors $1/(p^2+m^2)$ and $1/(q^2+m^2)$ are just the Green functions attached to the external legs: they do not enter the surviving integral. In momentum space we can "amputate" the diagram, which means omit the Green functions attached to the external legs. Note that the factorization (1.48) does not occur in coordinate space.

What remains is the "core" of our diagram, that is to say, its truly nontrivial part, which is, in the case at hand,

$$
\int \frac{d^4k}{(2\pi)^4}\frac{1}{k^2+m^2}. \quad (1.49)
$$

Unfortunately, the integral (1.49) is infinite, as are many integrals that we are going to work with. However, this kind of problem, which is the main topic of this book, does not concern us right now. What is important here is that we have identified a few tricks that can help us save a lot of effort, by working in momentum space and concentrating on what occurs *inside* the diagram, since what happens outside is not new. From a certain point onwards, we will not need to use double lines to denote the Green functions anymore, apart from the situations where it is really necessary: it will be understood that the internal lines carry Green functions, while the external lines do not.

Focusing on the cores of the diagrams, we can now formulate the Feynman rules that give the correlation functions (1.41) of a scalar field theory in arbitrary d dimensions, with arbitrary interactions, in momentum space.

The Fourier transform $\tilde{G}_{\mathrm{B}}(p)$ of the two-point function $\langle \varphi(x)\varphi(y) \rangle'$ is called *propagator*. We have

$$\langle \varphi(x)\varphi(y) \rangle' = \int \frac{\mathrm{d}^d p}{(2\pi)^d} e^{ip(x-y)} \tilde{G}_{\mathrm{B}}(p).$$

The propagator is graphically denoted by means of a line that connects two points. We associate a *vertex* with each interaction term of the Lagrangian. A vertex is graphically denoted by means of lines ending at the same point, also called *legs*. Each leg is a field φ. The value of the vertex is equal to minus the coefficient of the associated Lagrangian term, summed over the permutations of the identical legs. In momentum space, the momentum p of the Fourier transform $\tilde{\varphi}(p)$ is conventionally oriented towards the vertex.

For example, in the φ^4 theory we have (in arbitrary dimensions)

$$\underline{}_{p} = \frac{1}{p^2 + m^2} \qquad \times = -\lambda \tag{1.50}$$

Consider a correlation function (1.41) at $J = 0$ with $n = E$ external legs, and assume that we want to calculate its $\mathcal{O}(\lambda^k)$-corrections. To achieve this goal, we
1) assign a momentum p to every external leg, imposing overall momentum conservation;
2) draw all the diagrams G that have k vertices and E external legs;
3) assign a momentum q to every internal leg, imposing momentum conservation at every vertex.

Next, we associate an integral $\mathcal{I}_G$ with each diagram G as follows: we *a*) write the propagator associated with every internal leg;

b) multiply by the value of every vertex;

c) multiply by the combinatorial factor c_G explained below;

d) integrate over the surviving independent internal momenta q, with the measures $d^d q / (2\pi)^d$.

The combinatorial factor is given by the formula

$$c_G = \frac{s_G}{\prod_i n_i! \, c_i^{n_i}}. \tag{1.51}$$

Here, n_i is the number of vertices of type i contained in G, and c_i^{-1} is the combinatorial factor that multiplies the vertex of type i in the Lagrangian. For example, $c_i = N!$, if the ith vertex has N identical legs, such as φ^N. Instead, $c_i = N_1! N_2!$, if the vertex is $\varphi^{N_1} \varphi^{N_2}$. And so on. Finally, the numerator s_G is the number of Wick contractions that lead to the same diagram G.

The safest way to compute s_G is by drawing the vertices of G on a piece of paper, together with E points associated with the external legs. Then, we must count how many ways there are to connect the external legs to the legs attached to the vertices, and the legs of the vertices among themselves, to build the diagram G. The result of this counting is s_G. It is not advisable to avoid the counting just described, and compute s_G by means of shortcuts (typically based on the symmetry properties of the diagram, which may be difficult to spot), although some textbooks suggest to do so.

Normally, s_G is a huge number, to the extent that it almost simplifies the factors appearing in the denominator of (1.51). This is one reason why it is convenient to arrange the expansion in terms of diagrams. Nevertheless, sometimes it can be better, for theoretical purposes, to forget about the diagrams, and work directly on the sum over the Wick contractions, each of which has $s = 1$. In doing so, it is much easier to have control over the combinatorial factors. We will use this kind of expansion in some proofs later on.

The diagrams can also be classified according to the number L of their "loops". The precise definition of L is the number of independent internal momenta q, those on which we must integrate. Thus, formula (1.46) contains a one-loop diagram and a two-loop one.

We will see later that the expansion in powers of λ coincides with the expansion in the number of loops. Graphically, loops appear as closed internal lines. However, it is not always easy to count them as such.

Basically, the combinatorial factors are due to identical legs. This is the reason why, to simplify some formulas, it is common to divide each Lagrangian term by the number of permutations of its identical legs. For example, in the φ^4 theory we have multiplied the quadratic part of the Lagrangian by $1/2!$ and the vertex by $1/4!$. With different normalizations, the propagators and the vertices get multiplied by extra coefficients. Apart from that, the rules to construct the graphs and the formula for the combinatorial factors remain the same.

Finally, observe that the factors $1/(\prod_i n_i!)$ in c_G are brought by the expansion of the exponential in power series, e.g.,

$$\exp\left(-\frac{\lambda_4}{4!}\int\varphi^4 - \frac{\lambda_6}{6!}\int\varphi^6\right) = \sum_{n,m=0}^{\infty} \frac{(-\lambda_4)^n(-\lambda_6)^m}{n!m!(4!)^n(6!)^m}\left(\int\varphi^4\right)^n\left(\int\varphi^6\right)^m.$$

They correspond to the permutations of identical vertices.

We illustrate the calculation of combinatorial factors with a couple of examples. Consider the one-loop diagram of (1.46). It contains just one vertex with $c = 4!$. Moreover, we can easily verify that $s = 4 \cdot 3$, since the left external leg can be connected to the vertex in four ways, after which the right external leg can be connected to the vertex in three ways. In this particular case, the diagram is uniquely determined once the external legs are assigned. Thus, $c_G = (4 \cdot 3)/4! = 1/2$, which is indeed the factor that multiplies the diagram in formula (1.46), together with the value of the vertex, which is $-\lambda$.

Next, consider the diagram

$$(1.52)$$

It is made of three identical vertices, so we have a factor $1/(4!)^3$ and a factor $1/3!$. The coefficient s is equal to $3(4!)^3$. To calculate it, we first draw three vertices with four legs each, and four external points. Then we connect the points and the vertices in all the ways that lead to the graph we want. We begin from the up-left external leg, which can be attached to a vertex in $4 \cdot 3$ ways, where 3 is the number of vertices we can choose, and 4 is the number of legs of each vertex. Once that is done, the down-left external leg can be attached in just 3 ways, because the vertex it must be attached to is already determined. Next, the up-right external leg can be attached to vertices in $4 \cdot 2$ ways, after which the down-right external leg can be attached in 4 ways. At that point, consider an internal leg of the left vertex: it can be connected with another internal leg in 6 ways. When this is done, the remaining internal leg of the left vertex can be connected with 3 internal legs. Finally, the surviving internal legs can be connected in 2 ways. In total

$$c_G = \frac{4 \cdot 3 \cdot 3 \cdot 4 \cdot 2 \cdot 4 \cdot 6 \cdot 3 \cdot 2}{3!(4!)^3} = \frac{3(4!)^3}{3!(4!)^3} = \frac{1}{2}.$$

Because of (1.42), the information just given is also sufficient to determine the correlation functions (1.29) at $J = 0$. In particular, $Z(0) = \langle 1 \rangle' Z_0(0)$ is a sum over diagrams with no external legs. There is a simple way to characterize the correlation functions without primes. Indeed, they differ from the correlation functions with primes just because they do not receive contributions from the diagrams that contain subdiagrams with no external legs. This statement will be proved at the end of the next section. Here we just give a simple example: the two-point function without primes at $\mathcal{O}(\lambda)$ simply loses the last term of (1.46), so

$$\langle \varphi(x_1)\varphi(x_2) \rangle = G_B(x_1, x_2) - \frac{\lambda}{2} \int \mathrm{d}^4 y \, G_B(x_1, y) G_B(y, y) G_B(y, x_2) + \mathcal{O}(\lambda^2).$$

1.4 Generating functionals, Schwinger-Dyson equations

The rules given in the previous section determine the correlation functions with primes, and the generating functional $Z(J)$. It turns out that $Z(J)$ contains redundant information. For example, working with $W(J)$, instead of $Z(J)$, it is possible to reduce a lot of effort. A third functional, which is the Legendre transform of $W(J)$ with respect to J, and is denoted by $\Gamma(\Phi)$, allows us to further simplify the calculations.

In this section we study the generating functionals and their properties. We start by deriving a functional equation for $Z(J)$, called *Schwinger-Dyson equation*.

We begin by noting that the functional integral of a total functional derivative is zero. We have

$$0 = \int [\mathrm{d}\varphi] \frac{\delta}{\delta\varphi(x)} \exp\left(-S(\varphi) + \int J\varphi\right)$$

$$= \int [\mathrm{d}\varphi] \left(-\frac{\delta S}{\delta\varphi(x)} + J(x)\right) \exp\left(-S(\varphi) + \int J\varphi\right). \qquad (1.53)$$

Using the perturbative expansion, it is sufficient to prove this formula for the free field theory, with an arbitrary set of φ-insertions. Consider a massive field. In the discretized version of the functional integral, where we have a finite number of ordinary integrals, we obviously have the identity

$$0 = \int_{-\infty}^{+\infty} \prod_{\{i_\mu\}} [\mathrm{d}\varphi_{\{i_\mu\}}] \frac{\partial}{\partial\varphi_{\{k_\rho\}}} \left\{ \left(\prod_{\{n_\sigma\}\subset I} \varphi_{\{n_\sigma\}} \right) \exp\left(-S_{\mathrm{discr}}(\varphi_{\{i_\mu\}})\right) \right\},$$
$$(1.54)$$

for every $\{k_\rho\}$ and every set I of insertions $\varphi_{\{n_\sigma\}}$, where $S_{\mathrm{discr}}(\varphi_{\{i_\mu\}})$ is the free discretized action (1.24). Indeed, one integral, the one over $\varphi_{\{k_\rho\}}$, vanishes, because the exponential contains

$$\exp\left(-\frac{m^2}{2} \varphi_{\{k_\rho\}}^2\right),$$

which is sufficient to kill the contributions due to the boundary $\varphi_{\{k_\rho\}} \to \pm\infty$. Since (1.54) holds for every lattice space ℓ and size L, it also holds in the continuum limit.

The result is actually more general, to the extent that it also holds when the mass vanishes and the free-field action is not positive definite in the Euclidean framework (which is the case, among others, of gravity). Indeed, we should not forget that, although we are temporarily working in the Euclidean framework, the correct theory is the one in Minkowski spacetime. There, the functional integral (1.21) contains an oscillating integrand, which can be always damped at infinity by assuming that the field has a mass with a small negative imaginary part $-i\varepsilon$, which is later sent to zero. By doing so, we can prove that the identity (1.53) is always true in perturbative quantum field theory. The reader who is familiar with the operatorial formulation of quantum field theory will notice that this prescription is also the one that defines the correlation functions as T-ordered products. In other words, the functional integral automatically selects the time-ordered correlation functions.

Formula (1.53) gives

$$J(x)Z(J) = \int [d\varphi] \left[(-\Box + m^2)\varphi(x) + \frac{\lambda}{3!}\varphi^3(x) \right] \exp\left(-S(\varphi) + \int J\varphi \right),$$
$$(1.55)$$

which can be graphically represented as

$$(1.56)$$

Here the disc stands for $Z(J)$ and the dot stands for J. A leg attached to the disc is a functional derivative with respect to J, i.e., a φ-insertion. Three legs meeting at the same point x denote three functional derivatives with respect to $J(x)$.

To write (1.55), we have exchanged the functional integral with the derivatives contained in $\Box$. In general, we have the identity

$$\partial_\mu \langle \varphi(x) \cdots \rangle = \langle \partial_\mu \varphi(x) \cdots \rangle,$$

where the dots stand for arbitrary insertions at points different from x. We can prove this identity as follows. Consider the generating functional $Z(J)$ (1.23) and (for definiteness) the two-point function

$$\langle \varphi(x)\varphi(y) \rangle_J = Z(J)^{-1} \frac{\delta^2 Z(J)}{\delta J(x)\delta J(y)}.$$

If we write $J(x) = J_1(x) - \partial_\mu J_2^\mu(x)$ inside (1.23), where J_1 and J_2^μ are arbitrary, the functional derivative with respect to J_2^μ originates an insertion of $\partial_\mu \varphi(y)$. To see this, we must use

$$\int (J_1 - \partial_\mu J_2^\mu)\varphi = \int J_1 \varphi + \int J_2^\mu (\partial_\mu \varphi),$$

where the integration by parts can be justified by assuming that J_2^μ decreases rapidly enough at infinity. Indeed, since the sources J are test functions, we can choose them as smooth as we want and, if needed, with compact support. Thus, we find

$$\langle \varphi(x)\partial_\mu \varphi(y) \rangle_J = Z(J)^{-1} \frac{\delta^2 Z(J)}{\delta J_1(x)\delta J_2^\mu(y)} = Z(J)^{-1} \frac{\delta^2 Z(J_1 - \partial_\rho J_2^\rho)}{\delta J_1(x)\delta J_2^\mu(y)}$$

$$= Z(J)^{-1}\partial_\rho^{(y)}\frac{\delta^2 Z(J)}{\delta J(x)\delta J(y)} = \partial_\rho^{(y)}\langle\varphi(x)\varphi(y)\rangle_J.$$

Multiplying both sides of (1.55) by $G_B(y,x)$, integrating over x and relabeling $y \to x$, we obtain

$$\text{(1.57)}$$

where, as before, the double line stands for the Green function.

We can derive an alternative equation, that is to say,

$$(-\Box + m^2)\langle\varphi(x)\rangle_J = J(x) - \frac{\lambda}{3!}\langle\varphi^3(x)\rangle_J, \tag{1.58}$$

if we divide both sides of (1.55) by $Z(J)$. Again, if we insert the Green function in (1.58), we get

$$\langle\varphi(x)\rangle_J = \int \mathrm{d}^4y\, G_B(x,y)J(y) - \frac{\lambda}{3!}\int \mathrm{d}^4y G_B(x,y)\langle\varphi^3(y)\rangle_J. \tag{1.59}$$

Now, recalling that $Z = \exp(W)$, observe that

$$\langle\varphi^3\rangle_J = \mathrm{e}^{-W(J)}\frac{\delta^3}{\delta J^3}\mathrm{e}^{W(J)} = W''' + 3W'W'' + W'^3, \tag{1.60}$$

each apex denoting a J derivative. Then equation (1.59) can be graphically represented as

$$\text{(1.61)}$$

where now the disc denotes W. Again, the legs attached to the disc denote functional derivatives with respect to J.

The third generating functional $\Gamma(\Phi)$ is the Legendre transform of W. Define the functional $\Phi(J)$ as

$$\Phi(J)_x = \frac{\delta W(J)}{\delta J(x)} = \langle \varphi(x) \rangle_J. \tag{1.62}$$

From (1.28) we find

$$\Phi(J)_x = \int d^4y\, G_B(x,y) J(y) + \mathcal{O}(\lambda).$$

We can perturbatively invert $\Phi(J)$ and define the functional $J(\Phi)_x$ such that $J(\Phi(J))_x = 1$. We obtain

$$J(\Phi)_x = (-\Box + m^2)\Phi(x) + \mathcal{O}(\lambda). \tag{1.63}$$

Now, the functional $\Gamma(\Phi)$ is defined as

$$\Gamma(\Phi) = -W(J(\Phi)) + \int d^4x\, J(\Phi)_x \Phi(x). \tag{1.64}$$

We easily find

$$\Gamma(\Phi) = \frac{1}{2} \int d^4x \left((\partial_\mu \Phi)^2 + m^2 \Phi^2 \right) + \mathcal{O}(\lambda), \tag{1.65}$$

so Γ looks like a sort of "quantum action". In the literature it is often called *effective action*. Note, however, that in Minkowski spacetime Γ is not even real. The Γ functional collects the amplitudes that determine the S matrix.

Let us work out the Schwinger-Dyson equation satisfied by Γ. First observe that, since Γ is a Legendre transform, we have

$$\frac{\delta \Gamma(\Phi)}{\delta \Phi(x)} = J(\Phi)_x. \tag{1.66}$$

This relation can be easily verified by explicit differentiation. Second, using the formula for the derivative of the inverse function, we also find

$$\frac{\delta^2 W}{\delta J(x)\delta J(y)} = \frac{\delta \Phi(J)_y}{\delta J(x)} = \left(\frac{\delta J(\Phi)_x}{\delta \Phi(y)} \right)^{-1} = \left(\frac{\delta^2 \Gamma(\Phi)}{\delta \Phi(x)\delta \Phi(y)} \right)^{-1}. \tag{1.67}$$

We write this formula symbolically as $W_{xy} = 1/\Gamma_{xy}$, where the subscripts denote derivatives with respect to the arguments (J for W, Φ for Γ) at the specified points. Third,

$$W_{xyz} = -\int \frac{1}{\Gamma_{xs}} \frac{1}{\Gamma_{yt}} \frac{1}{\Gamma_{zu}} \Gamma_{stu}, \tag{1.68}$$

where the integral is over the repeated subscripts. Using (1.60)-(1.68), equation (1.58) becomes

$$-\frac{\delta \Gamma(\Phi)}{\delta \Phi(x)} = -(-\Box+m^2)\Phi(x) - \frac{\lambda}{3!}\left(\Phi^3(x) - \int \frac{1}{\Gamma_{xs}}\frac{1}{\Gamma_{xt}}\frac{1}{\Gamma_{xu}}\Gamma_{stu} + \frac{3}{\Gamma_{xx}}\Phi(x)\right).$$

$$(1.69)$$

Graphically, this formula reads

$$(1.70)$$

where the line with a cut denotes $1/\Gamma''$.

We know that the correlation functions $\langle \varphi \cdots \varphi \rangle$ can be expressed as functional derivatives of $Z(J)$ with respect to J, calculated at $J = 0$, and divided by $Z(0)$. Similarly, the functional derivatives of W with respect to J, calculated at $J = 0$, and the functional derivatives of Γ with respect to Φ, calculated at $\Phi = 0$, define W and Γ correlation functions, respectively. Our purpose is to characterize the correlation functions of Z, W and Γ more precisely, and work out the relations among them.

The functional Z is the generator of all the correlation functions. Instead, we prove that W is the generating functional of the *connected* correlation functions. That is to say, W contains precisely the contributions to Z that are originated by the connected diagrams. We write

$$W(J) = \sum_{n=0}^{\infty} \frac{1}{n!} \int \left(\prod_{i=1}^{n} d^4 x_i\right) \langle \varphi(x_1) \cdots \varphi(x_n) \rangle_c J(x_1) \cdots J(x_n),$$

where the subscript c stands for "connected".

Moreover, we prove that Γ is the generating functional of the connected, amputated (which means that the external legs carry no Green functions G_B) *one-particle irreducible* (commonly abbreviated as 1PI) correlation functions, which we simply call "irreducible". The irreducible diagrams are those that do not become disconnected by cutting one internal line. Precisely, we prove that $-\Gamma$ exactly contains the (amputated) contributions to Z and W that are due to irreducible diagrams, with only one exception: the free two-point function, which has an extra minus sign. We then write

$$-\Gamma(\Phi) = \sum_{n=0}^{\infty} \frac{1}{n!} \int \left(\prod_{i=1}^{n} d^4 x_i \right) \langle \varphi(x_1) \cdots \varphi(x_n) \rangle_{1PI} \Phi(x_1) \cdots \Phi(x_n).$$

To prove that the W and Γ correlation functions are connected and irreducible, respectively, it is sufficient to note that
$i)$ W and Γ are connected and irreducible, respectively, at the free-field level;
$ii)$ the W equation (1.61) and the Γ equation (1.70) are connected and irreducible, respectively;
$iii)$ equations (1.61) and (1.70) can be solved algorithmically from the free field theory.

Property $i)$ is obvious, from (1.28) and (1.65). We now prove that the equations (1.61) and (1.70) are connected and irreducible, respectively. Observe that equation (1.57), instead, is neither of the two. Indeed, (1.57) contains the product $J(x)Z(J)$, and generates disconnected contributions when we differentiate with respect to J. Equation (1.61) contains no products of functionals, which means that it is connected. On the other hand, it is clearly reducible. Finally, equation (1.70) is connected and irreducible. Indeed, the first three terms of (1.70) are the contributions to the classical field equations. Being local, they are just vertices, rather than diagrams, so they are irreducible. The other terms of (1.70) are clearly irreducible.

Next, we prove that the equations (1.57), (1.61) and (1.70) can be solved algorithmically, starting from the free field theory. Observe that by repeatedly differentiating those equations with respect to the sources, J or Φ, and later setting J or Φ to zero, we obtain relations among the correlation functions of Z, or W, or Γ. Each differentiation amounts to dropping a dot, as in the first term of (1.57) after the equal sign, or adding a leg to a disc, or blowing up an internal line into a three-leg disc, as per (1.67) and (1.68), and summing appropriately.

The right-hand sides of equations (1.57), (1.61) and (1.70) are the sums of two sets of contributions, which we call U_1 and U_2. The set U_1 is the one that does not carry a factor of λ. It contains no disc, or a disc with no leg. The set U_2 is the one that carries a factor of λ, and contains discs with at most three legs. An analogous decomposition holds for the differentiated equations, and is the crucial property to prove our construction. If we take n functional derivatives, the left-hand sides become discs with $n+1$ legs, which stand for the $(n+1)$-point correlation functions. The right-hand sides are, again, the sums of two types of contributions, U_1 and U_2. The set U_1 contains no factor of λ, and discs with at most n legs. In the cases of W and Γ such a set vanishes after a sufficient number of functional derivatives. The set U_2 contains a factor of λ and discs with at most $n+3$ legs.

Equations (1.57), (1.61) and (1.70) ensure that, if we want to determine the $(n+1)$-point function up to the order λ^k, it is sufficient to know the m-point functions, $m \leqslant n$, up to the order λ^k, and the m'-point functions, $m' \leqslant n+3$, up to the order λ^{k-1}. Iterating the argument r times, we find that, if we want to determine the $(n+1)$-point function up to the order λ^k, we need to know the m-point functions, $m \leqslant n-r+3h$, up to the orders λ^{k-h}, with $h = 0, 1, \dots r+1$. Taking $r = n + 3k$, we need to know the m-point functions, $m \leqslant 3(h-k)$, up to the order λ^{k-h}: if $h \neq k$ we have zero, if $h = k$ we have $Z_0(0)$, which can be normalized to 1. This proves that equations (1.57), (1.61) and (1.70) can be solved algorithmically, as claimed.

Strictly speaking, the arguments we have given apply to the correlation functions that contain at least one insertion. They do not allow us to reach $W(0)$ and $\Gamma(0)$, since the left-hand sides of (1.57), (1.61) and (1.70) are the derivatives of the functionals with respect to J, or Φ. Nevertheless, we can easily reach the derivatives of $W(0)$ and $\Gamma(0)$ with respect to any parameter we want. Consider, for example, the derivatives with respect to m^2. We have

$$\frac{\partial W(J)}{\partial m^2} = -\frac{1}{2} \int \mathrm{d}^4 x \left[\frac{\delta^2 W(J)}{\delta J(x)\delta J(x)} + \left(\frac{\delta W(J)}{\delta J(x)} \right)^2 \right]$$

$$= -\frac{\partial \Gamma(\Phi)}{\partial m^2} = -\frac{1}{2} \int \mathrm{d}^4 x \left[\frac{1}{\Gamma_{xx}} + \Phi(x)^2 \right].$$

The right-hand side of the first line is connected, while the right-hand side of the second line is irreducible (because the propagator $1/\Gamma_{xy}$ is evaluated at $y = x$). The same line of reasoning works for $\partial W(J)/\partial \lambda = -\partial \Gamma(\Phi)/\partial \lambda$. In the end, the derivative of W (Γ) with respect to any parameter is a collection of connected (irreducible) diagrams. What are really left out, in both $W(0)$ and $\Gamma(0)$, are just irrelevant additive constants.

We have considered, for simplicity, the φ^4 theory, but the results extend straightforwardly to any polynomial theory in arbitrary spacetime dimensions.

Clearly, the disconnected diagrams are products of connected ones, so $W(J)$ and $Z(J)$ contain the same amount of information. However, working with $W(J)$ instead of $Z(J)$ saves us some effort. In the free-field limit, for example, only the two-point function is connected, so $W(J)$ contains just one term (see (1.28)), while $Z(J)$ contains infinitely many, because it is the exponential of $W(J)$.

The simplification due to Γ is more clearly visible in momentum space, rather than in coordinate space. Observe that a convolution becomes a product after the Fourier transform. The reducible diagrams are those that can be split into two parts, connected by a single leg. In momentum space they factorize, so they "disconnect". Clearly, we lose no information if we concentrate on the "minimal" factors of such products. Working with Γ, we take advantage of this simplification.

So far, we have proved that all the diagrams that contribute to W (respectively, Γ) are connected (irreducible). We still have to prove the converse, i.e., that all the connected (irreducible) diagrams contribute to W (Γ). To show this, we proceed as follows.

Let us begin with W. Write

$$Z(J) = 1 + W(J) + \frac{1}{2!}W^2(J) + \frac{1}{3!}W^3(J)\cdots . \tag{1.71}$$

Since $Z(J)$ contains all the diagrams, and $W(J)$ contains only connected diagrams, $W(J)$ contains all the connected diagrams of $Z(J)$. Now, take the connected part of equation (1.71), and note that the powers $W^n(J)$, $n > 1$, can give only disconnected contributions. Using (1.43), we get

$$Z'(J)\big|_c = Z(J)\big|_c = 1 + W(J)\big|_c = 1 + W(J). \tag{1.72}$$

Thus, the connected diagrams contained in $Z'(J) = Z(J)$ and $W(J)$ coincide. Moreover, in these two functionals they appear with the same coefficients. This property ensures that the Feynman rules we have determined for Z can be used also for W: we just have to discard the diagrams that are disconnected.

Comparing the two sides of (1.71), we get, in the first few cases,

$$\langle\varphi(x)\rangle_c = \langle\varphi(x)\rangle, \qquad \langle\varphi(x)\varphi(y)\rangle_c = \langle\varphi(x)\varphi(y)\rangle - \langle\varphi(x)\rangle\langle\varphi(y)\rangle,$$

$$\langle\varphi(x)\varphi(y)\varphi(z)\rangle_c = \langle\varphi(x)\varphi(y)\varphi(z)\rangle - \langle\varphi(x)\varphi(y)\rangle\langle\varphi(z)\rangle - \langle\varphi(y)\varphi(z)\rangle\langle\varphi(x)\rangle$$
$$- \langle\varphi(z)\varphi(x)\rangle\langle\varphi(y)\rangle + 2\langle\varphi(x)\rangle\langle\varphi(y)\rangle\langle\varphi(z)\rangle.$$

Observe that $W(0)$ is the sum of the connected diagrams that have no external legs. Consider a correlation function $\langle\varphi(x_1)\cdots\varphi(x_n)\rangle$, $n > 0$, and write it in terms of W derivatives (see (1.60) for an example). It is easy to check that $W(0)$ never appears: only the derivatives $W^{(n)}$ with $n > 0$ are involved. Thus, the correlation function $\langle\varphi(x_1)\cdots\varphi(x_n)\rangle$ can be expressed as the sum of products of connected diagrams that have a nonvanishing number of external legs. This statement was left without proof at the end of the previous section. Instead, the correlation function $\langle\varphi(x_1)\cdots\varphi(x_n)\rangle' = \langle\varphi(x_1)\cdots\varphi(x_n)\rangle e^{W(0)-W_0(0)}$ contains products of all the connected diagrams, including those that have no external legs. The diagrams that appear in both correlation functions are multiplied by the same coefficients.

It remains to study the correlation functions of $-\Gamma$. From (1.66) and (1.70) we see that $J(\Phi)$ is a sum of irreducible diagrams. Consider (1.64) and restrict it to the irreducible diagrams. We have

$$-\Gamma(\Phi) = -\Gamma(\Phi)|_{1\text{PI}} = W(J(\Phi))|_{1\text{PI}} - \int J(\Phi)\Phi|_{1\text{PI}}. \qquad (1.73)$$

To manipulate this formula, it is convenient to write $J = (-\Box + m^2)\Phi + \Delta J$ and expand in powers of ΔJ, where $\Delta J = \mathcal{O}(\lambda)$ can be read from the right-hand sides of (1.69) and (1.70). We find

$$W((-\Box + m^2)\Phi) = W(J - \Delta J) = W(J) - \int \Delta J \frac{\delta W}{\delta J} + \frac{1}{2}\int \Delta J \frac{\delta^2 W}{\delta J \delta J}\Delta J + \cdots$$

Turning this expansion around, we can also write

$$W(J) - \int J\Phi = W((-\Box + m^2)\Phi) + \int(\Delta J - J)\frac{\delta W}{\delta J} - \frac{1}{2}\int \Delta J \frac{\delta^2 W}{\delta J \delta J}\Delta J$$
$$+ \cdots = W((-\Box + m^2)\Phi) - \int \Phi(-\Box + m^2)\Phi - \frac{1}{2}\int \Delta J \frac{\delta^2 W}{\delta J \delta J}\Delta J + \cdots$$

Now we take the one-particle irreducible contributions of both sides of this equation. Note that the last term, as well as the higher-order corrections collected inside the dots, always give reducible diagrams, since ΔJ contains vertices. Thus, we get

$$-\Gamma(\Phi) = W((-\Box + m^2)\Phi)\big|_{\text{1PI}} - \int \Phi(-\Box + m^2)\Phi. \qquad (1.74)$$

Replacing J by $(-\Box + m^2)\Phi$ inside $W(J)$ is equivalent to amputating the external legs, and attaching a field Φ to each of them. Formula (1.74) tells us that $-\Gamma$ contains the amputated irreducible diagrams of W, with exactly the same coefficient they have in W, apart from the free two-point function, which has an extra minus sign because of the last term of (1.74). Indeed, at the free-field level we have

$$\Gamma_0 = \frac{1}{2}\int \Phi(-\Box + m^2)\Phi = \frac{1}{2}\int J(-\Box + m^2)^{-1}J = W_0,$$

so $+\Gamma$ is the amputated W (instead of $-\Gamma$). Finally, the Feynman rules worked out for Z and W also work for $-\Gamma$ (apart from the free two-point function), provided we discard the reducible diagrams.

It is easy to see that the results of this section do not depend on the form of the vertex, nor the free-field action around which we perturb, nor the types of fields. For example, if we replace the interaction $\sim \int \varphi^4$ by $\sim \int \varphi^6$, or by the sum of $\int \varphi^4$ and $\int \varphi^6$, or even by interactions that contain derivatives, such as $\sim \int \varphi^2(\partial_\mu \varphi)^2$, etc., all the arguments given above can be generalized with obvious modifications. The only assumption that is crucial for the derivation is that the interactions be local, which means that each vertex should be the integral of a monomial built with the fields and their derivatives.

In the end, we find that in every local perturbative quantum field theory, the generating functional Z contains all the correlation functions, while W and Γ contain only the connected and the amputated, one-particle irreducible correlation functions, respectively. Moreover, the correlation functions appear in Z, W and Γ with the same coefficients, apart from the Γ free-field two-point function.

Exercise 1 *Integrating (1.70), calculate $\Gamma(\Phi)$ at the tree level and at one loop.*

Solution. The first line of (1.70) can be integrated straightforwardly, and gives $S(\Phi)$. The second line is made of two terms. The first of them can generate only two-loop diagrams, so we can neglect it. The second term gives diagrams that contain at least one loop. Thus, at the tree level the Γ functional coincides with the classical action: $\Gamma(\Phi) = S(\Phi)$.

To calculate the one-loop corrections it is sufficient to calculate Γ_{xx} at the tree level, which is just

$$S_{xy} \equiv \frac{\delta^2 S(\Phi)}{\delta\Phi(x)\delta\Phi(y)} = \left(-\Box + m^2 + \frac{\lambda}{2}\Phi^2(x)\right)\delta(x-y),$$

in the limit $y \to x$. Then we insert it into the last term of equation (1.70), which becomes

$$\frac{\lambda}{2}\Phi(x)\frac{1}{\Gamma_{xx}} = \frac{\lambda}{2}\Phi(x)\frac{1}{S_{xx}} = \frac{1}{2}\frac{\delta}{\Phi(x)}\int d^4y \ln S_{yz}|_{z\to y}, \qquad (1.75)$$

having used $\lambda\Phi = S'''$. Finally, the Γ functional reads

$$\Gamma(\Phi) = S(\Phi) + \frac{1}{2}\int d^4x \ln S_{xy}|_{y\to x} \equiv S(\Phi) + \frac{1}{2}\mathrm{tr}\left[\ln \frac{\delta^2 S(\Phi)}{\delta\Phi(x)\delta\Phi(y)}\right], \qquad (1.76)$$

plus two-loop corrections, plus unimportant constants. Although for clarity we have used the φ^4 theory to derive this result, it can be easily checked that formula (1.76) holds for an arbitrary action $S(\Phi)$, because the specific form of the action is actually not necessary for the derivation. $\Box$

The classical action $S(\varphi)$ and the functional $-W(J)$ satisfy an interesting duality relation. Consider iJ as the "fields", $S_J(iJ) \equiv -W(-iJ)$ as their classical action, and φ as the sources coupled to iJ. Then, the W functional is equal to $-S(\varphi)$ itself. Precisely,

$$\int [dJ] \exp\left(W(-iJ) + \int iJ\varphi\right) = \exp\left(-S(\varphi)\right). \qquad (1.77)$$

Indeed, using (1.23) the left-hand side can be written as

$$\int [dJd\varphi'] \exp\left(-S(\varphi') + i\int \varphi J - i\int \varphi' J\right).$$

Integrating over J we get the "functional δ function"

$$\delta_F(\varphi - \varphi') = \prod_x \delta(\varphi(x) - \varphi'(x)),$$

whose meaning can be easily understood from the discretized version of the functional integral. Finally, integrating over φ' we obtain the right-hand side of (1.77).

From the perturbative point of view, it does not matter whether J is multiplied by i or not. Thus, we can also write

$$\int [\mathrm{d}J] \exp\left(W(J) - \int J\varphi\right) = \exp\left(-S(\varphi)\right).$$

The meaning of this identity is that, if we take the diagrams that contribute to the connected correlation functions, replace their vertices by the connected diagrams themselves, and their propagators by minus the reciprocals of the two-point functions, and finally multiply by a minus sign for every external leg, the results we obtain are the vertices again, or, in the case of the two-point function, minus the reciprocal of the free propagator.

1.5 Advanced generating functionals

We can also define generating functionals for n-particle irreducible connected Green functions, that is to say, connected Green functions that become disconnected when n or fewer internal lines are cut. In this section we explain how. Although the new functionals are rarely met in the literature, they can help us gain a more complete picture of what we are doing. Moreover, certain generalizations of these functionals are useful to treat some topics of the next chapters.

We first study the generating functional of the two-particle irreducible Green functions. We introduce a new source $K(x, y)$ coupled to the bilinear $\varphi(x)\varphi(y)$, and define

$$Z(J, K) = \int [\mathrm{d}\varphi] \exp\left(-S(\varphi) + \int J\varphi + \frac{1}{2}\int \varphi K\varphi\right) = e^{W(J,K)},$$

where $\int \varphi K \varphi = \int \mathrm{d}x\, \varphi(x) K(x, y)\varphi(y)\,\mathrm{d}y$. Then, we define

$$\Phi(x) = \frac{\delta W}{\delta J(x)} = \langle \varphi(x) \rangle, \qquad N(x, y) = \frac{\delta^2 W}{\delta J(x)\delta J(y)} = \langle \varphi(x)\varphi(y)\rangle_c, \quad (1.78)$$

at nonzero J and K. Observe that

$$\frac{\delta W}{\delta K(x, y)} = \frac{1}{2}\left(N(x, y) + \Phi(x)\Phi(y)\right) = \frac{1}{2}\frac{\delta^2 W}{\delta J(x)\delta J(y)} + \frac{1}{2}\frac{\delta W}{\delta J(x)}\frac{\delta W}{\delta J(y)}.$$
$$(1.79)$$

This is a functional differential equation for $W(J, K)$. It shows that the K dependence is not unrelated to the J dependence, so the advanced functional $W(J, K)$ does not contain new information, but just the information already known, expressed in a different way.

Now, call $\Gamma_2(\Phi, N)$ the Legendre transform of $W(J, K)$ with respect to both J and K, that is to say,

$$\Gamma_2(\Phi, N) = -W(J, K) + \int \frac{\delta W}{\delta J} J + \int \frac{\delta W}{\delta K} K$$
$$= -W(J, K) + \int J\Phi + \frac{1}{2} \int (NK + \Phi K\Phi),$$

where NK stands for $N(x, y)K(x, y)$ and J and K are meant to be functions of Φ and N, obtained by inverting (1.78). It will be soon evident that this transform is indeed well defined. Differentiating Γ_2 we get

$$\frac{\delta \Gamma_2}{\delta \Phi(x)} = J(x) + \int K(x, y)\Phi(y)\mathrm{d}y, \qquad \frac{\delta \Gamma_2}{\delta N(x, y)} = \frac{1}{2} K(x, y). \qquad (1.80)$$

To retrieve $\Gamma(\Phi)$ from $\Gamma_2(\Phi, N)$ it is sufficient to set $K = 0$, because then $W(J, K)$ becomes precisely the functional $W(J)$ encountered before. Inverting (1.80) we obtain Φ and N as functions of J and K. Once K is set to zero, the relations $\Phi = \Phi(J, 0)$ and $N = N(J, 0)$ allow us to express J as a function $J(\Phi)$ of Φ, which coincides with the relation found in the previous sections, but also N as a function $N(\Phi)$ of Φ. Finally,

$$\Gamma(\Phi) = \Gamma_2(\Phi, N(\Phi)).$$

At $J = K = 0$ we have that Φ is the expectation value of the field and N is the full propagator.

If $\Gamma(\Phi, K)$ denotes the 1PI Γ functional associated with the modified classical action

$$S(\varphi, K) = S(\varphi) - \frac{1}{2} \int \varphi K \varphi, \qquad (1.81)$$

the functional $\Gamma_2(\Phi, N)$ can be seen as its Legendre transform

$$\Gamma_2(\Phi, N) = \Gamma(\Phi, K) + \frac{1}{2} \int K(N + \Phi\Phi) \qquad (1.82)$$

with respect to K.

Exercise 2 *Calculate* $\Gamma_2(\Phi, N)$ *for a free scalar field, and rederive* $\Gamma(\Phi)$.

Solution. The source $K(x,y)$ is like a nonlocal squared mass, so $W(J, K)$ can be formally obtained from the usual functional $W(J) = W(J, 0)$, by replacing the mass m^2 with $m^2 - K$. From (1.26) we get

$$W(J, K) = \frac{1}{2} \int J(-\Box + m^2 - K)^{-1}J - \frac{1}{2}\text{tr}\ln\left[-\Box + m^2 - K\right].$$

We immediately find

$$\Phi = (-\Box + m^2 - K)^{-1}J, \qquad N = (-\Box + m^2 - K)^{-1},$$

thus

$$\Gamma_2(\Phi, N) = \frac{1}{2} \int \left[(\partial_\mu \Phi)^2 + m^2\Phi^2\right] - \frac{1}{2}\text{tr}\ln N + \frac{1}{2}\text{tr}\left[(-\Box + m^2)N - 1\right].$$

Observe that objects such as $\ln N$ and N^{-1} are meaningful, since by (1.78) N^{-1} is just the propagator. Setting $K = 0$ we find $N = (-\Box + m^2)^{-1}$ and the usual free-field Γ-functional

$$\Gamma_2(\Phi, (-\Box + m^2)^{-1}) = \frac{1}{2} \int \left[(\partial_\mu \Phi)^2 + m^2\Phi^2\right] + \frac{1}{2}\text{tr}\ln(-\Box + m^2) = \Gamma(\Phi),$$

which agrees with (1.76). In an interacting theory we obtain this expression plus corrections proportional to the couplings.

Exercise 3 *Calculate* $\Gamma_2(\Phi, N)$ *at one loop for a generic theory* $S(\varphi)$.

Solution. We start from formula (1.76), which gives the most general one-loop Γ functional, and apply it to a classical theory with modified action (1.81). We obtain the one-loop Γ functional

$$\Gamma(\Phi, K) = S(\Phi) - \frac{1}{2} \int \Phi K \Phi + \frac{1}{2}\text{tr}\ln(S'' - K),$$

where S'' stands for S_{xy}. Now we further Legendre transform with respect to K. Differentiating, we obtain

$$\frac{\delta\Gamma}{\delta K} = -\frac{1}{2}\frac{1}{S'' - K} - \frac{1}{2}\Phi\Phi = -\frac{\delta W}{\delta K},$$

which gives

$$N = \frac{1}{S'' - K}. \tag{1.83}$$

Finally, using (1.82), the one-loop functional Γ_2 is

$$\Gamma_2(\Phi, N) = S(\Phi) - \frac{1}{2}\operatorname{tr}\ln N + \frac{1}{2}\operatorname{tr}\left[NS''(\Phi) - 1\right]. \tag{1.84}$$

$\square$

Now we study the diagrammatics of $\Gamma_2(\Phi, N)$. Since every (connected) one-loop diagram is two-particle reducible, unless it contains just one vertex (in which case it is called "tadpole"), it is useful to consider the difference $\tilde{\Gamma}_2(\Phi, N)$ between $\Gamma_2(\Phi, N)$ and its one-loop expression (1.84):

$$\tilde{\Gamma}_2(\Phi, N) = \Gamma_2(\Phi, N) - S(\Phi) + \frac{1}{2}\operatorname{tr}\ln N - \frac{1}{2}\operatorname{tr}\left[NS''(\Phi) - 1\right]. \tag{1.85}$$

Now, the functional $\Gamma(\Phi, K)$ is the set of 1PI diagrams of the theory $S(\varphi, K)$, namely, the set of 1PI diagrams of $S(\varphi)$ with inverse propagator shifted by $-K$. We separate the tree-level contribution $S(\Phi, K)$ of $\Gamma(\Phi, K)$ from the rest by writing

$$\Gamma(\Phi, K) = S(\Phi) - \frac{1}{2}\int \Phi K \Phi + \tilde{\Gamma}(\Phi, K). \tag{1.86}$$

The two-point function of $\Gamma(\Phi, K)$ is

$$\frac{\delta^2 \Gamma(\Phi, K)}{\delta \Phi \delta \Phi} = S''(\Phi) - K + \frac{\delta^2 \tilde{\Gamma}(\Phi, K)}{\delta \Phi \delta \Phi} = \left(\frac{\delta^2 W(J, K)}{\delta J \delta J}\right)^{-1} = \frac{1}{N}. \tag{1.87}$$

The last two equalities follow from (1.67) and the second formula of (1.78).

Now we switch to formula (1.82). Using (1.86) and (1.87), we find

$$\Gamma_2(\Phi, N) = S(\Phi) + \tilde{\Gamma}(\Phi, K) + \frac{1}{2}\operatorname{tr}\left[NS''(\Phi) + N\frac{\delta^2 \tilde{\Gamma}(\Phi, K)}{\delta \Phi \delta \Phi} - 1\right].$$

We must re-express K as a function of Φ and N on the left-hand side. We do so by considering the propagators and the vertices separately. By formula (1.87), all the propagators just become N. This means that, in the variables Φ and N, N is precisely *the internal line* of the diagrams. So, when we discuss reducibility, cutting an internal line is equivalent to taking a derivative with respect to N. Note that N does not vanish at $K = 0$, according to formula (1.83). We may say that N has a nonvanishing expectation value.

It remains to re-express the sources K that appear in the vertices. Observe that each K is attached to two φ legs, so two propagators N. Thus, we have to consider the product NKN. Using (1.87) we see that

$$NKN \rightarrow NS''(\Phi)N - N + N\frac{\delta^2\tilde{\Gamma}(\Phi, K)}{\delta\Phi\delta\Phi}N.$$

The sources K on the right-hand side can be treated recursively. Then, it is easy to see that the diagrams of $\Gamma_2(\Phi, N)$, and also those of $\tilde{\Gamma}_2(\Phi, N)$, are one-particle irreducible. The term $-(1/2)\text{tr} \ln N$ of $\Gamma_2(\Phi, N)$ can be considered one-particle irreducible as well, since, as noted above, N has a nonvanishing expectation value.

Working out the N derivative of $\tilde{\Gamma}_2$ and using (1.87), we get

$$\frac{\delta\tilde{\Gamma}_2}{\delta N} = \frac{\delta\Gamma_2}{\delta N} + \frac{1}{2N} - \frac{1}{2}S''(\Phi) = \frac{1}{2}\frac{\delta^2\tilde{\Gamma}(\Phi, K)}{\delta\Phi\delta\Phi}.$$

Repeating the argument above, we find that the diagrams of $\delta\tilde{\Gamma}_2/\delta N$ are also one-particle irreducible. Then the diagrams of $\tilde{\Gamma}_2$ are two-particle irreducible, because taking an N derivative is equivalent to cutting one internal line.

The functional Γ_∞ is defined by coupling sources $K_n(x_1, \ldots, x_n)$ to arbitrary strings $\varphi(x_1) \cdots \varphi(x_n)$ of φ insertions:

$$Z(J, K) = \int [\mathrm{d}\varphi] \exp\left(-S(\varphi) + \int J\varphi + \sum_{n=2}^{\infty} \frac{1}{n!} \int K_n \overbrace{\varphi \cdots \varphi}^{n}\right).$$

Then $W(J, K) = \ln Z(J, K)$, as usual, and

$$\Phi = \frac{\delta W}{\delta J} = \langle\varphi\rangle, \qquad N_n = \frac{\delta^n W}{\delta J \cdots \delta J} = \overbrace{\langle\varphi \cdots \varphi\rangle}^{n}_c. \qquad (1.88)$$

We have, in compact notation,

$$\frac{\delta W}{\delta K_n} = \frac{1}{n!}\overbrace{\langle\varphi \cdots \varphi\rangle}^{n} = \frac{1}{n!}\left.\mathrm{e}^{-W}\frac{\delta^n}{\delta J^n}\mathrm{e}^{W}\right|_{W' \rightarrow \Phi, W^{(k)} \rightarrow N_k}.$$

Performing the Legendre transform with respect to all the sources, we obtain the functional

$$\Gamma_\infty(\Phi, N) = -W(J, K) + \int \frac{\delta W}{\delta J}J + \sum_{n=2}^{\infty} \int \frac{\delta W}{\delta K_n}K_n,$$

where J and the sources K_n need to be expressed as functions of Φ and N_k, by inverting (1.88). The functional $\Gamma(\Phi)$ is retrieved by setting all the sources K_n to zero. The functional $\Gamma_2(\Phi, N)$ is obtained by setting all of them to zero but K_2, and so on.

1.6 Massive vector fields

So far, we have just considered scalar fields. Massive vector fields can be treated in a similar way, while fermions of spin 1/2 require that we extend the definition of functional integral to anticommuting variables. Finally, gauge fields need a separate treatment, since the definition of the functional integral in the presence of gauge symmetries is not straightforward, even in the Gaussian limit.

In the case of massive vector fields, we start from the free Proca action

$$S_{\text{free}}(W) = \int \mathrm{d}^4 x \left(\frac{1}{4} W_{\mu\nu}^2 + \frac{m^2}{2} W_\mu^2 \right), \tag{1.89}$$

where $W_{\mu\nu} \equiv \partial_\mu W_\nu - \partial_\nu W_\mu$. The field equations

$$-\Box W_\mu + \partial_\mu \partial_\nu W_\nu + m^2 W_\mu = 0 \tag{1.90}$$

ensure that the theory propagates only three degrees of freedom, at the classical level, since the divergence of (1.90) gives $m^2 \partial_\mu W_\mu = 0$. The propagator $G_{\mu\nu}(x, y) = \langle W_\mu(x) W_\nu(y) \rangle$ is the solution of the differential equation

$$(-\Box \delta_{\mu\nu} + \partial_\mu \partial_\nu + m^2 \delta_{\mu\nu}) G_{\nu\rho}(x, y) = \delta_{\mu\rho} \delta^{(4)}(x - y),$$

and can be easily expressed by means of the Green function G_{B} of the scalar field. Indeed, recalling (1.27), we find

$$G_{\mu\nu}(x, y) = \left(\delta_{\mu\nu} - \frac{\partial_\mu \partial_\nu}{m^2} \right) G_{\text{B}}(x, y) = \int \frac{\mathrm{d}^4 p}{(2\pi)^4} e^{ip \cdot (x-y)} \frac{\delta_{\mu\nu} + \frac{p_\mu p_\nu}{m^2}}{p^2 + m^2}. \tag{1.91}$$

At the quantum level, the degrees of freedom can be counted by counting the poles of the propagator in momentum space, after switching to Minkowski spacetime. Sticking to the Euclidean notation, the poles read $p^\mu = (\pm im, 0, 0, 0)$ in the particle rest frame. The numerator of the propagator, evaluated on a pole, is the matrix $\text{diag}(0, 1, 1, 1)$. The three positive eigenvalues are the propagating degrees of freedom, while the eigenvalue zero corresponds to the nonpropagating component $\partial_\mu W_\mu$.

When we add interactions, the Feynman rules and the diagrammatics follow straightforwardly, as well as the definitions of the generating functionals.

The fields of gauge theories are massless vectors. However, the massless limit of (1.91) is singular, which is why the gauge fields need a separate discussion. For the same reason, the ultraviolet limit of a theory that contains massive vectors is singular, because the mass becomes negligible at high energies. Another way to see the problem of massive vectors is that the propagator in momentum space behaves like $\sim p_\mu p_\nu/(m^2 p^2)$ for large p, instead of $\sim 1/p^2$. We will see that this behavior is not good enough for renormalizability. In general, an interacting quantum field theory that contains massive vector fields is nonrenormalizable. The same conclusion applies to the theories that contain massive fields of higher spins, which we do not treat here.

1.7   Fermions

The functional integral provides a formulation of quantum mechanics that is equivalent to the orthodox ones. Its main virtue is that it allows us to work with functions, instead of operators. In practice, summing over all the paths that connect the initial point to the final one has the same effect as working with objects that have nontrivial commutators. In some sense, the right-hand sides of the commutators $[\hat{q}, \hat{p}] = i$, $[\hat{a}, \hat{a}^\dagger] = 1$, where $[A, B] \equiv AB - BA$, are replaced by the functional integration.

We know that, to have consistency with the Fermi statistics in the operatorial approach, the second quantization of fermions is achieved by assuming that there exist annihilation and creation operators $\hat{a}_f$ and $\hat{a}_f^\dagger$ that satisfy the *anti*commutation relations $\{\hat{a}_f, \hat{a}_f^\dagger\} = 1$, $\{\hat{a}_f, \hat{a}_f\} = \{\hat{a}_f^\dagger, \hat{a}_f^\dagger\} = 0$, where $\{A, B\} = AB+BA$. We expect that the functional integral over the fermions can replace the right-hand side of the first anticommutator. We do not expect, however, that it can do more than that, for example allow us to work with commuting objects, instead of anticommuting ones. Indeed, the Pauli exclusion principle, which is the origin of the anticommutators, survives the classical limit $\hbar \to 0$. The right-hand sides of the commutators and the anticommutators vanish when $\hbar$ tends to zero, but the left-hand sides remain unchanged. This means that, in order to properly describe the fermions, we need to work with anticommuting objects, and define a suitable integral over them.

Such objects are called Grassmann variables. For the time being, we denote them by θ_i, $\bar{\theta}_i$. They satisfy

$$\{\theta_i, \theta_j\} = \{\theta_i, \bar{\theta}_j\} = \{\bar{\theta}_i, \bar{\theta}_j\} = 0.$$

We also need to define functions of such variables, then an "ordinary" integral over them, and finally the functional integral.

The definitions we work out below may sound a bit formal, at first. We have warned the reader that the mathematics must be upgraded in nontrivial ways, and possibly include notions that may sound unfamiliar. Quantum mechanics has already accustomed us to something like this, by teaching us how to work with quantities, such as the wave function, which do not have a direct connection with reality. We know that we just need to retrieve real numbers at the very end, when we compute the physical quantities. In quantum field theory, the situation gets much worse: most of the concepts we use in the intermediate steps have no direct connection with reality. Ultimately, we are free to introduce any objects we want, no matter how awkward they may look at first sight, as long as they are equipped with a set of consistent axioms that allow us to manipulate them, and are such that the predictions we obtain at the end make sense physically.

Consider a generic function of a single Grassmann variable θ. Making a Taylor expansion around $\theta = 0$, we find

$$f(\theta) = a + \theta b, \qquad a = f(0), \quad b = f'(0), \tag{1.92}$$

where a and b are constants. Every other term of the expansion disappears, since $\theta^2 = (1/2)\{\theta, \theta\} = 0$.

Similarly, a function of two variables θ, $\bar{\theta}$ reads

$$g(\theta, \bar{\theta}) = c + \theta d + \bar{\theta} e + \theta\bar{\theta} f,$$

c, d, e and f being other constants.

Ordinary commuting variables are called "c-numbers", to distinguish them from the Grassmann variables. If the function f of formula (1.92) is a *c*-number, then *a* is also a *c*-number, while *b* is an anticommuting constant. If *f* is anticommuting, then *a* is also anticommuting, while *b* is a *c*-number. We may say that *c*-numbers have bosonic statistics, while the Grassmann variables have fermionic statistics.

We introduce the differential $d\theta$, which is also anticommuting: $\{d\theta, \theta'\} = 0$ for every anticommuting θ'. Then the integral of $f(\theta)$ in $d\theta$ can be defined as follows, from the assumptions that it is linear and translational invariant. Linearity, i.e.,

$$\int d\theta \, f(\theta) = \left(\int d\theta \, 1\right) a + \left(\int d\theta \, \theta\right) b,$$

shows that it is sufficient to define the integrals of 1 and θ. Let us perform the change of variables $\theta = \theta' + \xi$, where ξ is constant and anticommuting. Translational invariance ensures $d\theta = d\theta'$, so

$$\int d\theta \, \theta = \int d\theta' \, (\theta' + \xi) = \int d\theta' \, \theta' + \left(\int d\theta' \, 1\right) \xi = \int d\theta \, \theta + \left(\int d\theta \, 1\right) \xi.$$

We conclude that the integral of 1 in $d\theta$ must vanish. Then, the integral of θ must not be zero, otherwise the integral would vanish identically. Normalizing it to 1, we have the formal rules

$$\int d\theta \, 1 = 0, \qquad \int d\theta \, \theta = 1,$$

which define the *Berezin integral*.

In practice, the Berezin integral behaves like a derivative. For example, under a rescaling $\theta' = c\theta$ we have

$$1 = \int d\theta' \, \theta' = c \int d(c\theta) \, \theta = \int d\theta \, \theta,$$

whence, differently from what happens with commuting variables,

$$d(c\theta) = \frac{1}{c} d\theta.$$

This rule coincides with the one of the derivative with respect to θ.

The basic Gaussian integral reads

$$\int d\bar{\theta} d\theta \, e^{-m\bar{\theta}\theta} = \int d\bar{\theta} d\theta \, (1 - m\bar{\theta}\theta) = m. \tag{1.93}$$

The minus sign disappears when we anticommute $\bar{\theta}$ with $d\theta$.

With more variables it is easy to prove that

$$\int \prod_{i=1}^{N} d\bar{\theta}_i d\theta_i \; \bar{\theta}_{i_1}\theta_{j_1} \cdots \bar{\theta}_{i_N}\theta_{j_N} = (-1)^N \varepsilon_{i_1\cdots i_N}\varepsilon_{j_1\cdots j_N}. \tag{1.94}$$

Indeed, the result must be completely antisymmetric in $i_1\cdots i_N$ and $j_1\cdots j_N$. Taking $i_k = j_k = k$ and using (1.93) we correctly get $(-1)^N$.

Then, defining the action

$$S(\bar{\theta},\theta) = \sum_{i,j=1}^{N} \bar{\theta}_i M_{ij}\theta_j,$$

where M_{ij} is some matrix, we get

$$\int \prod_{i=1}^{N} d\bar{\theta}_i d\theta_i \; e^{-S(\bar{\theta},\theta)} = \frac{(-1)^N}{N!} \int \prod_{i=1}^{N} d\bar{\theta}_i d\theta_i \; S^N(\bar{\theta},\theta)$$

$$= \frac{1}{N!}\varepsilon_{i_1\cdots i_N}\varepsilon_{j_1\cdots j_N} M_{i_1 j_1}\cdots M_{i_N j_N} = \det M. \tag{1.95}$$

Every other contribution coming from the exponential integrates to zero, because it cannot saturate the Grassmann variables θ and $\bar{\theta}$. We can easily generalize this formula to

$$Z(\bar{\xi},\xi) \equiv \int \prod_{i=1}^{N} d\bar{\theta}_i d\theta_i \; \exp\left(-S(\bar{\theta},\theta) + \sum_{i=1}^{N}(\bar{\xi}_i\theta_i + \bar{\theta}_i\xi_i)\right)$$

$$= \exp\left(\sum_{i,j=1}^{N} \bar{\xi}_i M_{ij}^{-1}\xi_j\right)\det M, \tag{1.96}$$

with the help of the translation $\theta = \theta' + M^{-1}\xi$, $\bar{\theta} = \bar{\theta}' + \bar{\xi}M^{-1}$.

Finally, a generic change of variables $\theta = \theta(\theta')$ gives the reciprocal of the usual Jacobian determinant,

$$\prod_{i=1}^{N} d\theta_i = \left(\det \frac{\partial\theta}{\partial\theta'}\right)^{-1} \prod_{i=1}^{N} d\theta'_i. \tag{1.97}$$

The derivatives with respect to the Grassmann variables can be placed to the left or to the right of the differentials $d\bar{\theta}$, $d\theta$. To keep track of what we are doing, it is convenient to define left- and right-derivatives, ∂_l and ∂_r, which differ at most by a minus sign. Precisely, the differential of a function can be written as

$$\mathrm{d}f(\bar{\theta},\theta) = \frac{\partial_r f}{\partial\bar{\theta}_i}\mathrm{d}\bar{\theta}_i + \frac{\partial_r f}{\partial\theta_i}\mathrm{d}\theta_i = \mathrm{d}\bar{\theta}_i\frac{\partial_l f}{\partial\bar{\theta}_i} + \mathrm{d}\theta_i\frac{\partial_l f}{\partial\theta_i}.$$

If B is a bosonic function and χ is a bosonic (fermionic) variable, the identity $\partial_l B/\partial\chi = \partial_r B/\partial\chi$ $(\partial_l B/\partial\chi = -\partial_r B/\partial\chi)$ holds. If F is a fermionic function, the identity $\partial_l F/\partial\chi = \partial_r F/\partial\chi$ holds no matter the statistics of χ.

Of course, $\partial_l/\partial\bar{\theta}$ and $\partial_l/\partial\theta$ are anticommuting objects, as well as $\partial_r/\partial\bar{\theta}$ and $\partial_r/\partial\theta$. But observe that

$$\frac{\partial_r}{\partial\bar{\theta}_i}\frac{\partial_l}{\partial\theta_j} = \frac{\partial_l}{\partial\theta_j}\frac{\partial_r}{\partial\bar{\theta}_i}.$$

We can define averages

$$\langle\theta_{i_1}\cdots\theta_{i_n}\bar{\theta}_{j_1}\cdots\bar{\theta}_{j_n}\rangle = \frac{1}{Z(\bar{\xi},\xi)}\frac{\partial_l}{\partial\bar{\xi}_{i_1}}\cdots\frac{\partial_l}{\partial\bar{\xi}_{i_n}}\frac{\partial_r}{\partial\xi_{j_n}}\cdots\frac{\partial_r}{\partial\xi_{j_1}}Z(\bar{\xi},\xi)\Bigg|_{\xi=\bar{\xi}=0}.$$

In particular,

$$\langle\theta_i\bar{\theta}_j\rangle = M_{ij}^{-1}, \qquad \langle\theta_i\theta_j\bar{\theta}_k\bar{\theta}_l\rangle = M_{il}^{-1}M_{jk}^{-1} - M_{ik}^{-1}M_{jl}^{-1}. \tag{1.98}$$

We can also have integrals over both commuting variables x and anticommuting variables θ. Writing $\bar{z} = (\bar{x},\bar{\theta})$ and $z = (x,\theta)$, we define the *superdeterminant* as

$$(\mathrm{sdet}M)^{-1} \equiv \frac{1}{(2\pi)^N}\int\mathrm{d}\bar{z}\mathrm{d}z\exp(-\bar{z}^t M z)$$

$$= \frac{1}{(2\pi)^N}\int\mathrm{d}\bar{x}\mathrm{d}x\mathrm{d}\bar{\theta}\mathrm{d}\theta\exp(-\bar{x}^t A x - \bar{x}^t B\theta - \bar{\theta}C x - \bar{\theta}D\theta),$$

where the normalization factor is chosen so that $\mathrm{sdet}\mathbb{1} = 1$,

$$M = \begin{pmatrix} A & B \\ C & D \end{pmatrix}$$

and each block A, B, C and D is a $N \times N$ matrix, where A, D contain commuting entries and B, C contain anticommuting entries.

To compute the superdeterminant, we perform the translations $\bar{y}^t = \bar{x}^t + \bar{\theta} C A^{-1}$ and $y = x + A^{-1} B\theta$, and observe that in the variables $\bar{\zeta} = (\bar{y}, \bar{\theta})$, $\zeta = (y, \theta)$, we have $\bar{z}^t M z = \bar{y}^t A y + \bar{\theta}(D - CA^{-1}B)\theta$, so the integrals over commuting and anticommuting variables factorize. At the end, we find

$$\text{sdet}\, M = \frac{\det A}{\det(D - CA^{-1}B)}. \tag{1.99}$$

A useful property, which we do not prove here, is

$$\ln \text{sdet}(\exp M) = \text{str} M \equiv \text{tr} A - \text{tr} D, \tag{1.100}$$

where "str" denotes the so-called supertrace. Moreover, the infinitesimal variation

$$\delta \text{sdet} M = \delta \exp(\text{str} \ln M) = (\text{sdet} M)\text{str}(M^{-1}\delta M) \tag{1.101}$$

holds. In this book we just need a special case of this formula, when $M = 1 + \delta M$ and δM is small:

$$\text{sdet}(1 + \delta M) \sim 1 + \text{tr}[\delta A] - \text{tr}[\delta D] = 1 + \text{str}[\delta M]. \tag{1.102}$$

This result can be proved by expanding formula (1.99) to the first order in δM.

Finally, a generic change of variables $(\bar{z}, z) \to (\bar{\zeta}, \zeta)$ gives

$$d\bar{z}dz = d\bar{\zeta}d\zeta \; \text{sdet}\frac{\partial(\bar{z}, z)}{\partial(\bar{\zeta}, \zeta)}.$$

Again, we leave this formula without proof, but it is easy to derive the infinitesimal version that we need later. For $(\bar{z}, z) = (\bar{\zeta} + \delta\bar{\zeta}, \zeta + \delta\zeta)$ we find, to the first order,

$$d\bar{z}dz \sim d\bar{\zeta}d\zeta \left(1 + \text{str}\frac{\partial(\delta\bar{\zeta}, \delta\zeta)}{\partial(\bar{\zeta}, \zeta)}\right). \tag{1.103}$$

The minus sign inside the supertrace is due to the exponent -1 of the Jacobian determinant in (1.97).

The continuum limit is now straightforward. Consider for example free Dirac fermions, which have the action

$$S(\bar{\psi}, \psi) = \int d^4x \; \bar{\psi}(\partial\!\!\!/ + m)\psi. \tag{1.104}$$

Here $\partial\!\!\!/ = \gamma^\mu \partial_\mu$ and γ^μ are the γ matrices in Euclidean space, which satisfy $\{\gamma^\mu, \gamma^\nu\} = 2\delta^{\mu\nu}$ and $(\gamma^\mu)^\dagger = \gamma^\mu$. The Green function $G_F(x, y) = \langle \psi(x)\bar\psi(y)\rangle$ is the solution of the equation

$$(\partial\!\!\!/_x + m)G_F(x, y) = \delta^{(4)}(x - y).$$

We find

$$G_F(x, y) = (-\partial\!\!\!/_x + m)G_B(x, y) = \int \frac{\mathrm{d}^4 p}{(2\pi)^4} \frac{-i p\!\!\!/ + m}{p^2 + m^2} e^{ip\cdot(x-y)}. \qquad (1.105)$$

The generating functionals are defined as

$$Z(\bar\xi, \xi) = \int [\mathrm{d}\bar\psi \mathrm{d}\psi] \exp\left(-S(\bar\psi, \psi) + \int \bar\xi\psi + \int \bar\psi\xi\right) = e^{W[\bar\xi,\xi]},$$

where $\int \bar\xi\psi$ and $\int \bar\psi\xi$ stand for $\int \mathrm{d}^4 x\, \bar\xi(x)\psi(x)$ and $\int \mathrm{d}^4 x\, \bar\psi(x)\xi(x)$, respectively. Using (1.96) we obtain

$$W(\bar\xi, \xi) = \int \mathrm{d}^4 x\, \bar\xi(x)G_F(x, y)\xi(y)\mathrm{d}^4 y$$

plus an irrelevant constant.

Wick's theorem reads

$$\langle \chi_1 \cdots \chi_{2n}\rangle = \sum_P \varepsilon_P \langle \chi_{P(1)}\chi_{P(2)}\rangle \cdots \langle \chi_{P(2n-1)}\chi_{P(2n)}\rangle,$$

where χ_i can be either $\psi(x_i)$ or $\bar\psi(x_i)$, while ε_P is the signature of the permutation P. Precisely, ε_P is equal to 1 or -1, depending on whether $\{P(1), P(2), \cdots, P(2n)\}$ is obtained from $\{1, 2, \cdots, 2n\}$ by means of an even or odd number of permutations of two nearby elements. The free correlation functions with an odd number of insertions vanish.

We can build the perturbative expansion around the free theory by following the guidelines we learned in the case of scalar fields. Let us consider, for definiteness, the four-fermion model

$$S_4(\bar\psi, \psi) = \int \mathrm{d}^4 x \left(\bar\psi(\partial\!\!\!/ + m)\psi - \frac{\lambda}{4}(\bar\psi\psi)^2\right). \qquad (1.106)$$

The Feynman rules are

$$\overrightarrow{\underset{\alpha \quad p \quad \beta}{}} = \left(\frac{1}{i\not{p}+m}\right)_{\alpha\beta} \qquad\qquad = \frac{\lambda}{2}(\delta^{\alpha\delta}\delta^{\beta\gamma} - \delta^{\alpha\gamma}\delta^{\beta\delta})$$

with legs labeled α (top left), δ (top right), β (bottom left), γ (bottom right).

$$\tag{1.107}$$

where α, β, etc. are spinor indices. Observe that:
1) the vertex and the Lagrangian term have opposite signs;
2) each incoming line of the vertex is a ψ and each outgoing line is a $\bar\psi$;
3) the fermion lines are drawn with arrows pointing from the right to the left; then their spinor indices are ordered from the left to the right;
4) if the Lagrangian term is ordered by putting each $\bar\psi$ to the left and each ψ to the right, the vertex is drawn by putting the $\bar\psi$ legs to the left, and the ψ legs to the right;
5) if we order the fields $\bar\psi$ (respectively, ψ) from the left to the right, the legs associated with them are ordered from the top to the bottom (resp., from the bottom to the top);
6) the vertices must include all the permutations of identical lines, each permutation carrying a factor -1.

    Point 1) is due to the minus sign that appears in front of the action in e^{-S}. Point 6) is why the factor 4 of $\lambda/4$ drops out.

    It is convenient to draw the vertices with legs of variable lengths, so that, if we associate a dot with each external endpoint:
7) from the left to the right, the dots are ordered according to the positions of $\bar\psi$ and ψ in the Lagrangian vertex, as in (1.107).

    A diagram G is built as follows.
8) Draw the vertices one after another, horizontally (in no particular order), so that the dots of one vertex are all located before, or after, the dots of every other vertex; let R_V denote the row made by the vertices.
9) Draw an $\times$ for each $\bar\psi$ external leg to the left of R_V, and an $\times$ for each ψ external leg to the right of R_V, ordering them according to what explained in the points 4,5,7) above.
10) Connect the vertices among themselves, and to the external legs, to build the internal and the external lines of G:

a) to build an internal G line, move a ψ dot to the left of the $\bar{\psi}$ dot it must be connected to, and multiply by a factor -1 every time the ψ dot crosses another dot (with respect to the horizontal ordering); connect the two dots and then suppress them;

b) to build an external G line, move a ψ dot, or a $\bar{\psi}$ dot, to the $\times$ it must be connected to, and multiply by a factor -1 for every dot crossing; connect the dot to the $\times$, and then suppress the dot.

In simple theories, where the vertices contain at most one $\bar{\psi}$ and one ψ, the rules 8-10) can be replaced by a shortcut, that is to say,

11) every fermion loop is multiplied by a factor -1.

Finally, in evaluating the diagram, remember that it must be multiplied by its own combinatorial factor and, because of 3),

12) the spinor indices are ordered according to the reversed arrows.

If the diagram G is irreducible, and is calculated according to the rules just explained, it contributes to the correlation functions of $-\Gamma$, the external legs being arranged according to the rules 4,5,7). The rules 8-10) are the graphical translation of Wick's theorem, once we expand as in (1.44). The minus sign of point 11) can be quickly explained by means of the Berezin integral. Consider, for example,

$$\int \prod_i d\bar{\theta}_i d\theta_i \, (\bar{\theta} V_1 \theta)(\bar{\theta} V_2 \theta) \exp\left(-\bar{\theta}^t M \theta\right),$$

where the V_i's are matrices that appear in the vertices, possibly depending on other fields. Using (1.98) we obtain

$$-\mathrm{tr}[V_1 M^{-1} V_2 M^{-1}] + (-\mathrm{tr}[V_1 M^{-1}])(-\mathrm{tr}[V_2 M^{-1}]).$$

The first contribution corresponds to a one-loop diagram that contains both vertices, and is indeed multiplied by -1. The second contribution is the product of two diagrams, each of which has one loop and contains a single vertex.

The result is easily generalized to the diagrams with an arbitrary number of loops. The minus sign appears when we move the first $\bar{\theta}$ to the far right:

$$\prod_{i=1}^{N} (\bar{\theta} V_i \theta) \rightarrow -\mathrm{tr}[V_1 \langle \theta\bar{\theta}\rangle V_2 \langle \theta\bar{\theta}\rangle V_3 \langle \theta\bar{\theta}\rangle V_4 \theta \cdots V_N \langle \theta\bar{\theta}\rangle].$$

The simplest example of interacting scalar-fermion theory is the (massless) Yukawa model

$$S(\varphi, \psi) = \int d^4x \left(\frac{1}{2}(\partial_\mu \varphi)^2 + \bar{\psi}(\not{\partial} + g\varphi)\psi + \frac{\lambda}{4!}\varphi^4 \right) \tag{1.108}$$

with Feynman rules

$$\tag{1.109}$$

The rules to build the diagrams are the same as before.

The functional Γ is defined as the Legendre transform

$$\Gamma(\Phi, \bar{\Psi}, \Psi) = -W(J, \bar{\xi}, \xi) + \int J\Phi + \int \bar{\xi}\Psi + \int \bar{\Psi}\xi,$$

where

$$\Phi = \frac{\delta W}{\delta J}, \qquad \Psi = \frac{\delta_l W}{\delta \bar{\xi}}, \qquad \bar{\Psi} = \frac{\delta_r W}{\delta \xi}.$$

All the arguments applied before, to prove that W and Γ are the generating functionals of the connected and amputated, one-particle irreducible diagrams, respectively, can be repeated here with obvious modifications. Actually, the derivation can be extended to the most general local perturbative quantum field theory. The Feynman rules for Z, W and Γ are the same, since the diagrams appear in each functional with the same coefficients (apart from the free Γ two-point function).

Working on Γ makes the study of renormalization much simpler. For this reason, from now on we mostly concentrate on the irreducible diagrams.

1.8 Locality

The locality assumption, which is crucial in perturbative quantum field theory, has intriguing aspects. It requires that the action $S(\varphi, \bar{\psi}, \psi, V_\mu, \ldots)$ be a local functional of the fields. It should be noted, however, that the action

S does not contain the true interactions, which are encoded into the correlation functions. As we will see, the correlation functions are most of the times nonlocal. So, why should we require that the classical action be local? Even more, why should we require that there exist a classical action, and the theory be built on it? Why not investigate all the conceivable Γ functionals, directly?

An attempt like this has been made, decades ago, but did not lead to substantial progress. The point is that, if we do not have a sufficiently constrained starting point, such as a local (and renormalizable, as we will see) classical action, what we can say is so arbitrary that making predictions becomes almost impossible. As we know from quantum mechanics, exploring the microscopic world is not like exploring the classical world. We can make only sporadic experiments, and just collect data here and there. The macroscopic objects emit a huge, practically infinite, number of photons, which are collected by our eyes, or instruments, in a finite amount of time. Each photon is like an individual experiment, so we gather an infinite amount of information at once. Which is why, classically, we do not worry so much about constraining the physical laws *a priori*: the experimental observation is so powerful that it constrains them for us *a posteriori*. In quantum field theory, on the other hand, we would not go very far, if we did not have ways to select the theories and their interactions a priori.

All this is fine, but prompts a dilemma: why should nature arrange itself so as to make us capable of investigating it? And isn't it a twisted assumption to require that the observable interactions be built starting from a local "classical" action, which may have no direct connection with the experimental observation of the classical world?

Well, this recipe, together with the other recipes we identify in the rest of the book, is what remains of the old correspondence principle. We call S the "classical action" not because it has something to do with the classical phenomena, but because it is the starting point of a process of *quantization*. Since we cannot have a direct intuition of the quantum world, the best we can hope for is to be able to quant-*ize* a phantom of the classical world. If we did not even have this possibility, we would probably not know how to make progress in high-energy physics.

After properly formulating local, renormalizable, perturbative quantum field theory, we will be ready to explore more general quantum field theories, including the nonrenormalizable and the nonlocal ones. What we stress here is that if we make a conceptual jump that is too far reaching at the very beginning, we risk plunging into the domain of absolute arbitrariness. We have to start from what is working for sure, or has more chances to work, and depart from that little by little.

Chapter 2

Renormalization

We have seen that the perturbative expansion generates ill defined integrals, such as (1.49). This is the first serious problem of the "creative approximation" we undertook in the previous chapter. It is useful to compare the situation we are facing with the problem of giving sense to an improper integral over the real line,

$$\int_{-\infty}^{+\infty} dx \ f(x). \tag{2.1}$$

Written like this, this expression has no intrinsic meaning, and needs to be defined. A natural suggestion, due to Riemann, is to view it as the limit

$$\lim_{\Lambda \to \infty} \int_{-\Lambda}^{+\Lambda} dx \ f(x). \tag{2.2}$$

Precisely, a "cutoff" Λ is inserted, to turn the original integral (2.1) into a definite one. After calculating the definite integral, the limit $\Lambda \to \infty$ is studied. If the limit exists, the integral is said to be convergent. If the limit does not exist, the integral is said to be divergent.

In quantum field theory, we do not have to define *one* integral, but a *theory*, which contains an infinite number of integrals, one for each diagram. Different diagrams are related to one another by various identities. The physical quantities involve, in general, sums, products and convolutions of integrals. If a single integral does not converge, the reason may simply be that we have isolated it from the rest of the theory in an inconvenient way. This happens, for example, when the "divergence" disappears by changing the variables (fields, spacetime coordinates, or momenta, couplings, and any other parameter of the theory), i.e., by performing all sorts of operations that normally do not change the physics. When that is the case, the divergence is not a problem, but just a blunder due to an unfortunate parametrization of the theory.

Before jumping to the conclusion that the theory is ill defined, and maybe throw it away, we take advantage of the freedom we have. Instead of requiring that the $\Lambda \to \infty$ limit make sense integral by integral, right after inserting the cutoff, we concentrate on the physical quantities. While Λ is still finite, we take the liberty of performing a number of operations that are expected to be innocuous (but will they really be so?). That is to say, we move the Λ divergences around, from one quantity to another, by performing changes of field variables and reparametrizations, hoping to make the physical quantities convergent for $\Lambda \to \infty$.

We can summarize this attitude by saying that, instead of viewing each integral as an improper integral, we view the whole theory as an "improper theory". Then the goal is to answer the following question: is there a rearrangement, based on reparametrizations and field redefinitions, after which the theory (which means: the set of physical quantities built with the theory) has a convergent $\Lambda \to \infty$ limit?

The insertion of a cutoff is called *regularization*. The rearrangement of the Λ divergences that allows us to achieve the goal just stated is called *renormalization*. Of course, we will have to prove that the physical results do not depend on the way we regularize and renormalize the theory.

The cutoff is a useful tool to classify the divergences. In principle, we do not strictly need to introduce one. In the literature there exist several regularization-independent approaches that do not make explicit uses of cutoffs. On the other hand, working with a cutoff is very convenient, because it helps us keep track of what we do when we move the divergences around. The goal of the rearrangement is to identify "the right places for the divergences", so that, after moving the "infinities" to their destinations, the limit $\Lambda \to \infty$ makes sense *in all the physical quantities*, but not necessarily in the single integrals and the quantities that are mathematically useful to build the theory, but physically meaningless. If this program works, we obtain a consistent (perturbative) definition of the local quantum field theory.

Definition 1 *A theory is called convergent if, possibly after a reparametrization, all the physical quantities are convergent in the limit $\Lambda \to \infty$. Other-*

wise it is called divergent.

The definition of convergent theory is not equivalent to the definition of "renormalizable" theory. We will appreciate the difference later.

The cutoff (2.2) is the simplest and most intuitive way to smooth out the singularities. It amounts to state that the domain of integration is bounded to momenta that have a modulus smaller than Λ. Clearly, this trick makes every integral convergent at finite Λ. For example, the two-point function $G_B(x, y)$ is divergent at coincident points. At finite Λ we find

$$G_B(x, x) = \int_{|p| \leqslant \Lambda} \frac{d^4 p}{(2\pi)^4} \frac{1}{p^2 + m^2} = \frac{1}{16\pi^2} \left[\Lambda^2 - m^2 \ln \left(1 + \frac{\Lambda^2}{m^2} \right) \right]$$

$$= \frac{1}{16\pi^2} \left[\Lambda^2 - m^2 \ln \frac{\Lambda^2}{m^2} + m^2 \mathcal{O} \left(\frac{m^2}{\Lambda^2} \right) \right]. \tag{2.3}$$

When Λ is sent to infinity, we have a quadratic divergence, which is the term proportional to Λ^2, plus a logarithmic divergence, which is the term proportional to $\ln \Lambda$, plus finite contributions.

The divergences occur at large momenta, or, equivalently, coincident points. They are basically due to the locality of our theories. If we admit nonlocal, rather than pointlike, interactions, then we can easily build theories with no divergences. However, as we have remarked at the end of the previous chapter, the nonlocalities open the door to a huge arbitrariness. It is better to first deal with the divergences in local theories, then investigate the nonlocal theories. Besides, we have already said that the divergences of isolated integrals are not the true problem: it would be a mistake to throw away theories just because they look divergent at first sight.

Definition 2 *Given a theory having Feynman rules F, a regularization is any deformation F_Λ of the Feynman rules that gives sense to all the individual integrals generated by the perturbative expansion, and is such that F_Λ gives back F when the deformation is switched off.*

We stress that the regularization does not need to be physical, because the cutoff must be eventually removed. Actually, the most common regularization techniques are unphysical, in the sense that the regularized theories are not physically acceptable as quantum field theories *per se*, typically because they violate some physical principles. The cutoff is an example of unphysical regularization, since it violates unitarity. Indeed, it excludes the contributions of high frequencies from the integrals, while unitarity says (loosely speaking) that the set of particles that circulate in a loop must coincide with the set of ingoing and outgoing particles.

On the other hand, the violation of locality does not sound like the violation of a physical principle, so a theory that is regularized in a nonlocal way might well be physical in its own right. Yet, we stress again that the intrinsic arbitrariness of nonlocal theories makes us postpone their investigation as physical theories to the very end. For the moment, the problems we find in local theories are rather welcome, because they may provide us with selective criteria to isolate the theories that should be discarded from the theories that can be accepted. If the selection is powerful enough, we might be able to make predictions that are worth of being tested in experiments.

It may be objected that inserting a cutoff *à la* Riemann may not be the smartest choice. Luckily, there exists a regularization technique that meets the needs of perturbative quantum field theory in a much more economic and efficient way. This is the dimensional regularization.

2.1   Dimensional regularization

The *dimensional regularization* is a regularization technique that is based on the continuation of the spacetime dimension to complex values. As awkward as this concept may sound at first, we recall that we just need to provide a consistent set of axioms, and a formal construction, to make the manipulations we need, and generate physical predictions that make sense. Everything in between can be as artificial as we wish.

Consider an integral $\mathcal{I}_4$ in four dimensions, in momentum space. Denote the integrated momentum by p and the external momenta by k. Assume that the integrand is Lorentz invariant in Minkowski spacetime, and a rational function. To dodge a number of nuisances that are not important for the present discussion, we continue to work in the Euclidean framework. There, the integrand is invariant under rotations, and can be expressed as a function f of p^2 and the scalar products $p \cdot k$:

$$\mathcal{I}_4(k) = \int \frac{\mathrm{d}^4 p}{(2\pi)^4} f(p^2, p \cdot k).$$

An *analytic integral* $\mathcal{I}_D(k)$ in complex D dimensions can be associated with $\mathcal{I}_4$ as follows. Replace the four-dimensional integration measure $\mathrm{d}^4 p$ with a formal D-dimensional measure $\mathrm{d}^D p$, and include a $(2\pi)^D$ in the denominator for convenience, instead of $(2\pi)^4$. Replace p_μ and k_μ with formal D-dimensional vectors inside the integrand. This gives

$$\mathcal{I}_D(k) = \int \frac{\mathrm{d}^D p}{(2\pi)^D} f(p^2, p \cdot k). \tag{2.4}$$

We want to define the analytic integral in D dimensions so that it coincides with the ordinary integral $\mathcal{I}_d(k)$ when D takes integer values d and $\mathcal{I}_d(k)$ is convergent. When $\mathcal{I}_d(k)$ is not convergent, we want to use $\mathcal{I}_D(k)$ to classify its divergence.

To achieve this goal, we start by writing the analytic integral $\mathcal{I}_D(k)$ in spherical coordinates. The measure reads

$$\int \mathrm{d}^D p = \int_0^\infty p^{D-1} \mathrm{d}p \times$$

$$\times \int_0^{2\pi} \mathrm{d}\theta_1 \int_0^\pi \mathrm{d}\theta_2 \sin\theta_2 \cdots \int_0^\pi \mathrm{d}\theta_{D-1} \sin^{D-2}\theta_{D-1},$$

any time D is integer. When L is integer and greater than one, we also have

$$\int_0^{2\pi} \mathrm{d}\theta_1 \int_0^\pi \mathrm{d}\theta_2 \sin\theta_2 \cdots \int_0^\pi \mathrm{d}\theta_{L-1} \sin^{L-2}\theta_{L-1} \, 1 = \frac{2\pi^{L/2}}{\Gamma\left(\frac{L}{2}\right)},$$

which is the total solid angle in L dimensions.

Since the external momenta k are finitely many, because a Feynman diagram has a finite number of external legs, the integrand of (2.4) depends on finitely many angles $\theta_{D-L}, \cdots, \theta_{D-1}$. The number D is still unspecified and, for the time being, we can imagine that it is integer and sufficiently large, in any case larger than L. Then we can write

$$\mathcal{I}_D(k) = \frac{1}{2^{D-1}\pi^{(D+L)/2}\Gamma\left(\frac{D-L}{2}\right)} \times$$

$$\times \int_0^\infty \mathrm{d}p \int_0^\pi \mathrm{d}\theta_{D-L} \int_0^\pi \mathrm{d}\theta_2 \cdots \int_0^\pi \mathrm{d}\theta_{D-1} p^{D-1} \bar{f}(p, \theta_{D-L} \cdots \theta_{D-1}, D). \tag{2.5}$$

The function $\bar{f}$ includes the factors $\sin^{i-1}\theta_i$, $i = D - L, \ldots D - 1$.

Now, the expression on the right-hand side of (2.5) is meaningful for generic complex D. Assume that there is an open domain $\mathcal{D}$ in the complex plane where the integral $\mathcal{I}_D(k)$, written as in (2.5), is well-defined. Evaluate $\mathcal{I}_D(k)$ in $\mathcal{D}$. Then, analytically continue the function $\mathcal{I}_D(k)$ from $\mathcal{D}$ to the rest of the complex plane. The value of this function at $D = 4$, if it exists, is the physical value of the integral $\mathcal{I}_4(k)$. If it does not exist, the function $\mathcal{I}_D(k)$ has poles around $D = 4$. Such poles classify its divergences.

For example,

$$\mathcal{I}_D(m) \equiv \int \frac{\mathrm{d}^D p}{(2\pi)^D} \frac{1}{p^2 + m^2} = \frac{1}{2^{D-1}\pi^{D/2}\Gamma\left(\frac{D}{2}\right)} \int_0^\infty \mathrm{d}p \frac{p^{D-1}}{p^2 + m^2}. \qquad (2.6)$$

The integral is well-defined in the strip $0 < \mathrm{Re}\, D < 2$. The analytic continuation gives the result (see Appendix A, formula (A.5))

$$\frac{\Gamma\left(1 - \frac{D}{2}\right) m^{D-2}}{(4\pi)^{D/2}} = \frac{1}{16\pi^2}\left[-\frac{2m^2}{\varepsilon} + m^2\left(\ln\frac{m^2}{4\pi} - 1 + \gamma_E\right) + \mathcal{O}(\varepsilon)\right], \qquad (2.7)$$

where $\gamma_E = 0.5772...$ is the Euler-Mascheroni constant. The right-hand side of formula (2.7) is the expansion around four dimensions, having written $D = 4 - \varepsilon$ and used formula (A.8).

Observe that the term $m^2 \ln m^2$ coincides with the one of (2.3). The logarithmic divergences of (2.3) and (2.7) coincide after identifying $\ln \Lambda$ with $1/\varepsilon$. We can compare these two types of divergences by noting that

$$\int_{|p|\geqslant\delta} \frac{\mathrm{d}^D p}{(2\pi)^D (p^2)^2} \sim \frac{1}{8\pi^2\varepsilon} + \text{finite}, \qquad \int_{\delta\leqslant|p|\leqslant\Lambda} \frac{\mathrm{d}^4 p}{(2\pi)^4 (p^2)^2} \sim \frac{1}{8\pi^2}\ln\Lambda,$$

where δ is an infrared cutoff. The other contributions to (2.3) and (2.7) differ from each other. In particular, (2.7) contains no analogue of the quadratic divergence Λ^2. Differences and similarities will become clear later.

What happens when the integral, expressed in the form (2.5) does not admit a domain of convergence $\mathcal{D}$? Or when it admits more disconnected domains of convergence?

First, observe that the Feynman rules of a local quantum field theory can only give rational integrands. Then, if the domain of convergence $\mathcal{D}$ exists, it is always unique (a strip $X < \mathrm{Re}\, D < Y$), which ensures that the analytic continuation is also unique, as well as the value of the integral in D dimensions. The situation where an integral admits two disconnected convergence domains cannot occur.

If an integral does not admit a convergence domain, assume that we can decompose the integrand f into a finite sum of integrands f_i, such that each of them admits its own convergence domain $\mathcal{D}_i$. Then we define the integral of f as the sum of the integrals of each f_i. For example, the integrand $f \equiv 1$ does not admit a domain of convergence. However, writing

$$1 = \frac{p^2 + m^2}{p^2 + m^2} = f_1 + f_2, \qquad f_1 = \frac{p^2}{p^2 + m^2}, \qquad f_2 = \frac{m^2}{p^2 + m^2},$$

we see that f_1 and f_2 admit the domains of convergence $-2 < \operatorname{Re} D < 0$ and $0 < \operatorname{Re} D < 2$, respectively. We thus find

$$\int \frac{\mathrm{d}^D p}{(2\pi)^D} f_1 = \frac{Dm^D \Gamma\left(-\frac{D}{2}\right)}{2^{D+1}\pi^{D/2}}, \qquad \int \frac{\mathrm{d}^D p}{(2\pi)^D} f_2 = m^2 \mathcal{I}_D(m).$$

Summing the two contributions, we discover that the analytic integral of one is actually zero. The same integral, treated with the cutoff method, behaves like Λ^4. We learn that the dimensional regularization kills every powerlike divergence. It is sensitive only to the logarithmic divergences, which manifest themselves as poles in ε.

With exactly the same procedure we can calculate the analytic integral of $(p^2)^\alpha$, for every complex α: we find again 0. More generally, let $f(p)$ be a rational function of p. Let α_{IR} and α_{UV} denote the exponents such that

$$f(p) \sim (p^2)^{\alpha_{\mathrm{IR}}}, \qquad f(p) \sim (p^2)^{\alpha_{\mathrm{UV}}},$$

for $p \to 0$ and $p \to \infty$, respectively. Decompose the integrand as

$$f(p)\left(\frac{p^2 + m^2}{p^2 + m^2}\right)^n = \sum_{k=0}^{n} \binom{n}{k} (m^2)^{n-k} \frac{f(p)(p^2)^k}{(p^2 + m^2)^n}.$$

The integral of the k-th term of the sum is convergent in the strip $-2\alpha_{\mathrm{IR}} - 2k < \operatorname{Re} D < 2n - 2\alpha_{\mathrm{UV}} - 2k$, which is nontrivial if its width $2n - 2\alpha_{\mathrm{UV}} + 2\alpha_{\mathrm{IR}}$ is strictly positive. Note that the width is k independent. Thus, if we choose n sufficiently large, in particular larger than $\alpha_{\mathrm{UV}} - \alpha_{\mathrm{IR}}$, all the terms of the sum can be integrated.

Concluding, we can always decompose the analytic integral of a rational function as a finite sum of integrals admitting nontrivial convergence domains. The construction easily extends to multiple integrals. Since a local quantum field theory can only generate rational integrands, these arguments prove that the dimensional-regularization technique is able to define every integral we need.

It remains to prove that our definition is consistent. We do not provide a complete proof here, but collect the basic arguments and mention the key properties of the integral.

First, the analytic integral is linear, and invariant under translations and rotations. In particular, the result does not depend on the center of the polar coordinates used to write (2.5). Moreover, the usual formulas for the multiple integration and the change of variables hold.

The rules of multiple integration deserve some comment. It is always safe to split an analytic integral in D dimensions as the sequence of two analytic integrals in D_1 and D_2 dimensions, with $D = D_1 + D_2$, which are defined as explained above:

$$\int \frac{\mathrm{d}^D p}{(2\pi)^D} = \int \frac{\mathrm{d}^{D_1} p_1}{(2\pi)^{D_2}} \int \frac{\mathrm{d}^{D_2} p_2}{(2\pi)^{D_2}}.$$

Sometimes, however, it is convenient to split the integral as an analytic integral followed by an ordinary integral. For example,

$$\int \frac{\mathrm{d}p_1}{2\pi} \int \frac{\mathrm{d}^{D-1} p_2}{(2\pi)^{D-1}}, \qquad \int \frac{\mathrm{d}^4 p_1}{(2\pi)^2} \int \frac{\mathrm{d}^{-\varepsilon} p_2}{(2\pi)^{-\varepsilon}}, \tag{2.8}$$

and so on. This kind of decomposition also works. However, the outside integral is still to be meant in the analytic sense. Precisely, after evaluating the inside integral, we obtain the ordinary integral of a function f that depends on D. That integral must be evaluated in a domain $\mathcal{D}$ where it converges, and analytically continued to the rest of the complex plane, as explained above. If a domain $\mathcal{D}$ does not exist, it must be written as a finite linear combination of ordinary integrals that separately admit domains of convergence $\mathcal{D}_i$. For example, if we use the second split of (2.8) on $\mathcal{I}_D(m)$, we can represent it as a four-dimensional integral:

$$\mathcal{I}_D(m) = \frac{\Gamma\left(1 + \varepsilon/2\right)}{(4\pi)^{-\varepsilon/2}} \int \frac{\mathrm{d}^4 p_1}{(2\pi)^2} \frac{1}{(p_1^2 + m^2)^{1+\varepsilon/2}}.$$

Neglecting the prefactor, which tends to 1 when ε tends to zero, this formula can be viewed as an alternative regularization of the integral. It does not change the integration per se and does not introduce a cutoff for the large momenta. Instead, it replaces the propagator with

$$\frac{1}{(p^2 + m^2)^{1+\varepsilon/2}},$$

where ε is a complex number. These integrals have to be calculated in a complex domain of ε values where they converge, and then analytically continued to the rest of the complex plane. In the literature, this procedure is known as *analytic regularization*. The good feature of the analytic regularization is that it deals with ordinary integrals all the time, so its consistency is easier to prove. We anticipate that, however, it breaks gauge invariance, while the dimensional regularization manifestly preserves it. Using the analytic regularization (or the cutoff one), gauge invariance has to be recovered by hand, which is possible, but requires a lot of effort. The dimensional regularization is a sort of rationalized analytic regularization, which knows how to rearrange itself so as to preserve gauge invariance at no cost.

Finally, it is normally not safe to split an analytic integral as an ordinary integral followed by an analytic integral, e.g.

$$\int \frac{\mathrm{d}^{-\varepsilon}p_1}{(2\pi)^{-\varepsilon}} \int \frac{\mathrm{d}^4 p_2}{(2\pi)^2}$$

because the ordinary integral might not converge. Check it on $\mathcal{I}_D(m)$.

2.1.1 Limits and other operations in D dimensions

The limits can be evaluated along similar guidelines. Consider a function $f(D, x)$. Its limit $f(D, x_0)$ for $x \to x_0$ is defined by applying the following two rules:

a) search for an open set $\mathcal{D}$ of the complex plane where the limit exists, calculate it there, and analytically continue the result to the complex plane;

b) if $f(D, x)$ admits no such $\mathcal{D}$, search for a decomposition of $f(D, x)$ into a finite sum $\sum_i f_i(D, x)$, such that each $f_i(D, x)$ admits a complex domain $\mathcal{D}_i$ where the limit exists, proceed as in point a) for each $f_i(D, x)$ and sum the analytic continuations $f_i(D, x_0)$.

As an example, consider the integral

$$\int \frac{\mathrm{d}^D p}{(2\pi)^D} \frac{\Lambda^2}{(p^2 + m^2)(p^2 + m^2 + \Lambda^2)}. \tag{2.9}$$

It can be evaluated by means of formula (A.2) of Appendix A, which allows us to express it as

$$\int_0^1 \mathrm{d}x \int \frac{\mathrm{d}^D p}{(2\pi)^D} \frac{\Lambda^2}{(p^2 + m^2 + x\Lambda^2)^2}.$$

Then formula (A.4) gives

$$\Lambda^2 \frac{\Gamma\left(2 - \frac{D}{2}\right)}{(4\pi)^{D/2}} \int_0^1 \mathrm{d}x\, (m^2 + x\Lambda^2)^{D/2-2}$$
$$= \frac{\Gamma\left(1 - \frac{D}{2}\right) m^{D-2}}{(4\pi)^{D/2}} \left[1 - \left(1 + \frac{\Lambda^2}{m^2}\right)^{D/2-1}\right]. \tag{2.10}$$

If we take Λ to infinity in the integrand of (2.9) we get (2.7). Now, consider the final result (2.10). It admits a regular limit only in the domain $\mathrm{Re}\,D < 2$. The analytic continuation of the limit in such a domain gives again (2.7).

To interchange derivatives and integrals, derivatives and limits, and perform all sorts of similar operations, we must follow the same guidelines, namely: *a*) decompose the function f into a finite sum of functions f_i each of which admits a domain $\mathcal{D}_i$ of the complex plane where the operation can be performed ordinarily, once the integral is expressed in the form (2.5), *b*) analytically continue each result to the complex plane, and *c*) sum the analytic continuations.

2.1.2   Functional integration measure

Now we prove an important property that is going to be useful in many contexts. We say that a function of the fields and their derivatives, evaluated at the same spacetime point, is ultralocal if it depends on the derivatives of the fields up to a finite order. It does not need to be polynomial in the fields and their derivatives. We prove that

Theorem 1 *using the dimensional-regularization technique, the functional integration measure is invariant under every ultralocal change of field variables.*

Proof. Let φ^i denote the fields and $\varphi^i \to \varphi^{i\prime}$ the change of field variables. If the field redefinition is ultralocal, there exists a finite number of ultralocal functions $F_{ij}^{\mu_1\cdots\mu_n}$ such that

$$\frac{\delta\varphi^{i\prime}(x)}{\delta\varphi^j(y)} = \sum_{n=0}^N F_{ij}^{\mu_1\cdots\mu_n}(\varphi(x))\partial_{\mu_1}\cdots\partial_{\mu_n}\delta^{(D)}(x-y), \tag{2.11}$$

and the Jacobian determinant can be written as

$$\mathcal{J} = \det\frac{\delta\varphi^{i\prime}(x)}{\delta\varphi^j(y)} = \exp\left(\mathrm{tr}\,\frac{\delta\varphi^{i\prime}(x)}{\delta\varphi^j(y)}\right) = \exp\left(\int \mathrm{d}^D x \frac{\delta\varphi^{i\prime}(x)}{\delta\varphi^i(x)}\right)$$

$$= \exp\left(\sum_{n=0}^{N} \partial_{\mu_1} \cdots \partial_{\mu_n} \delta^{(D)}(0) \int d^D x \, F_{ij}^{\mu_1 \cdots \mu_n} (\varphi(x)) \right).$$

Because of (2.11), the exponent is a finite sum of local functionals multiplied by $\delta^{(D)}(0)$, or derivatives of $\delta^{(D)}(0)$. Such expressions vanish using the dimensional regularization, because in momentum space they read

$$\partial_{\mu_1} \cdots \partial_{\mu_n} \delta^{(D)}(0) = i^n \int \frac{d^D p}{(2\pi)^D} p_{\mu_1} \cdots p_{\mu_n}. \tag{2.12}$$

Recalling that the analytic integral is invariant under rotations, we obtain zero when n is odd, but also zero when n is even. Indeed,

$$\int \frac{d^D p}{(2\pi)^D} p_{\mu_1} \cdots p_{\mu_{2k}} \propto (\delta_{\mu_1 \mu_2} \cdots \delta_{\mu_{2k-1} \mu_{2k}} + \text{perms.}) \int \frac{d^D p}{(2\pi)^D} (p^2)^k = 0. \tag{2.13}$$

□

The theorem we just proved is very general. It also holds when the change of variables depends on derivatives of arbitrary degrees, but can be treated as a perturbative series of ultralocal terms. Moreover, it holds for all types of fields: scalars, fermions, vectors, tensors, as well as fields of higher spins. To include fields of different statistics in the proof, it is sufficient to replace the determinant with the superdeterminant and the trace with the supertrace.

We say that a function is perturbatively local if it can be expanded perturbatively as a series of ultralocal polynomials in the fields and their derivatives, evaluated at the same spacetime point. In some situations we may just use the term "local" in this extended sense.

2.1.3 Dimensional regularization for vectors and fermions

In the context of the dimensional regularization, the coordinates x^μ, the momenta p_μ, the Kronecker tensor $\delta_{\mu\nu}$, and so on, have to be viewed as purely formal objects. We need to give a consistent set of operations to manipulate such objects, so that the correct four dimensional results are retrieved when D tends to 4. Similarly, the vector fields A_μ, the gamma matrices γ_μ and the spinors ψ^α also have to be considered as formal objects. In particular, the gamma "matrices" should not be viewed as true matrices, although we keep calling them with their usual name.

We define the D-dimensional Dirac algebra starting from a set of formal objects γ_μ that are equipped with a formal trace operation, and satisfy the following axioms:

$$\{\gamma_\mu, \gamma_\nu\} = 2\delta_{\mu\nu}\mathbb{1}, \qquad \gamma_\mu^\dagger = \gamma_\mu, \qquad \mathrm{tr}[\gamma_{\mu_1} \cdots \gamma_{\mu_{2n+1}}] = 0,$$
$$\mathrm{tr}[AB] = \mathrm{tr}[BA], \qquad \mathrm{tr}[\mathbb{1}] = f(D), \qquad f(4) = 4. \qquad (2.14)$$

In particular, the formal trace is cyclic and vanishes on an odd product of gamma matrices. Using the formal Dirac anticommutation relations, that is to say, the first axiom of (2.14), we can reduce every trace to the trace of the identity, which we call $f(D)$. The function $f(D)$ must be equal to 4 in four dimensions, but is otherwise arbitrary.

Specifically, the axioms (2.14) imply

$$\mathrm{tr}[\gamma_{\mu_1} \cdots \gamma_{\mu_{2n}}] = \sum_{i=2}^{2n} (-1)^i \delta_{\mu_1 \mu_i} \mathrm{tr}[\gamma_{\mu_2} \cdots \hat{\gamma}_{\mu_i} \cdots \gamma_{\mu_{2n}}], \qquad (2.15)$$

where $\hat{\gamma}_{\mu_i}$ means that the matrix γ_{μ_i} is dropped. The proof is identical to the one in four dimensions. In particular,

$$\mathrm{tr}[\gamma_\mu \gamma_\nu] = f(D)\delta_{\mu\nu},$$
$$\mathrm{tr}[\gamma_\mu \gamma_\nu \gamma_\rho \gamma_\sigma] = f(D)\left(\delta_{\mu\nu}\delta_{\rho\sigma} - \delta_{\mu\rho}\delta_{\nu\sigma} + \delta_{\mu\sigma}\delta_{\nu\rho}\right).$$

We also have the identities

$$\gamma_\mu \gamma_\mu = D\mathbb{1}, \qquad \gamma_\mu \gamma_\rho \gamma_\mu = (2 - D)\gamma_\rho.$$

It seems that in D dimensions everything proceeds smoothly, with minor modifications with respect to the usual formulas. This is not completely true. In four dimensions we can also define a matrix γ_5 that satisfies $\{\gamma_\mu, \gamma_5\} = 0$. A matrix with such properties does not exist in complex D dimensions. This problem is related to the appearance of an important "anomaly". Another object that cannot be extended to D dimensions is the Levi-Civita tensor $\varepsilon_{\mu\nu\rho\sigma}$, because it would have a complex number of indices. For the moment, we ignore these difficulties, and limit ourselves to parity invariant, nonchiral theories, where γ_5 and $\varepsilon_{\mu\nu\rho\sigma}$ do not appear in the Lagrangian and the Feynman diagrams. Another fact that is worth mentioning is that in odd dimensions it can be inconsistent to assume that the trace of an odd product of gamma matrices vanishes. For example, in three dimensions the trace

$\text{tr}[\sigma_i\sigma_j\sigma_k]$, where σ_i are the Pauli matrices, is not zero, but proportional to the Levi-Civita tensor ε_{ijk}. There exist modified versions of the dimensional regularization that bypass these difficulties, but we will not mention them in this book.

The dimensionally regularized versions of the models studied so far have formally identical Feynman rules (1.50), (1.107) and (1.109). However, for $D \neq 4$ the couplings are dimensionful even when they are dimensionless in $D = 4$. It is convenient to redefine them in a dimensionless way, by isolating suitable powers of an energy scale μ. For example, the Lagrangians (1.38) and (1.108) become

$$S(\varphi) = \int \mathrm{d}^D x \left(\frac{1}{2}(\partial_\mu\varphi)^2 + \frac{m^2}{2}\varphi^2 + \frac{\lambda\mu^\varepsilon}{4!}\varphi^4 \right) \tag{2.16}$$

and

$$S(\varphi, \psi) = \int \mathrm{d}^D x \left(\frac{1}{2}(\partial_\mu\varphi)^2 + \bar{\psi}\left(\partial\!\!\!/ + g\mu^{\varepsilon/2}\varphi\right)\psi + \frac{\lambda\mu^\varepsilon}{4!}\varphi^4 \right), \tag{2.17}$$

respectively. In the new parametrization, both g and λ are dimensionless in arbitrary D, and the Feynman rules are (1.50) and (1.109) with the replacements $g \to g\mu^{\varepsilon/2}$ and $\lambda \to \lambda\mu^\varepsilon$.

2.2 Divergences and counterterms

Now that we know that each diagram is associated with a well regularized integral, we can study the general properties of the diagrammatics.

Consider a diagram G with V vertices, E external legs and I internal legs. Assign an independent momentum to each internal and external leg. In total, this gives $I + E$ momenta. Once we impose the momentum conservation in each vertex, we remain with $I + E - V$ independent momenta. Observe that the external legs contain $E - 1$ independent momenta, because the E-th momentum is related to the other ones by the global momentum conservation. Therefore, the diagram G contains $I + E - V - (E - 1) = I - V + 1 = L$ independent internal momenta. The integral associated with G is performed over those momenta, and L is called the number of *loops* of the diagram. The "topological" formula

$$L - I + V = 1 \tag{2.18}$$

holds for every diagram, in every theory. It is called topological, because it coincides with Euler's formula for simple polyhedra, namely,

$$v - e + f = 2,$$

where v is the number of vertices, e is the number of edges and f is the number of faces of the polyhedron. The correspondence with (2.18) is $v = V$, $e = I$ and $f = L + 1$. Indeed, dropping the external legs and adding the "loop at infinity", which is the $(L+1)$-th face, a graph becomes a generalized polyhedron (which still satisfies Euler's formula).

Another very general fact is that the expansion in the number of loops coincides with the expansion in powers of $\hbar$. Although we have set $\hbar = 1$ so far, we can easily restore the $\hbar$ dependence by writing the generating functionals $Z(J)$ and $W(J)$ as

$$Z(J) = \int [\mathrm{d}\varphi]\, \exp\left(-\frac{1}{\hbar}S(\varphi) + \int J\varphi\right) = \exp\left(\frac{1}{\hbar}W(J)\right),$$

while $\Gamma(\Phi)$ is defined as before. In the new Feynman rules a propagator gets a factor $\hbar$ and a vertex gets a factor $1/\hbar$. Therefore, each diagram is multiplied by a factor

$$\hbar^{I-V} = \frac{\hbar^L}{\hbar},$$

having used (2.18). The diagrams contribute to Z in the usual way. If they are connected they contribute to $W/\hbar$, because $Z = \exp(W/\hbar)$. If they are irreducible they contribute to $-\Gamma/\hbar$. We thus see that the L-loop contributions to W and Γ are multiplied by $\hbar^L$.

Consider the "φ_d^N theory", that is to say, the d-dimensional scalar field theory with interaction φ^N. Its action reads

$$S(\varphi) = \int \mathrm{d}^d x \left(\frac{1}{2}(\partial_\mu \varphi)^2 + \frac{m^2}{2}\varphi^2 + \lambda\frac{\varphi^N}{N!}\right). \tag{2.19}$$

For the moment, we do not need to continue the physical dimension to complex values. Let $[\mathcal{O}]$ denote the dimension of an object $\mathcal{O}$ in units of mass. Coordinates have dimension -1, while momenta have dimension 1. Since the action is dimensionless, the Lagrangian must have dimension d.

From the kinetic term, or the mass term, we can read the dimension of φ. Then, we can read the dimension of λ from the vertex. We find

$$[x] = -1, \qquad [\partial] = 1, \qquad [\varphi] = \frac{d}{2} - 1, \qquad [\lambda] = N\left(1 - \frac{d}{2}\right) + d. \quad (2.20)$$

Consider again a diagram G with V vertices, E external legs and I internal legs. Since N legs are attached to each vertex, we have NV legs in total. Of these, E exit the diagram and $2I$, connected in pairs, build the internal legs, each of which is attached to two vertices. Therefore, the identity

$$E + 2I = NV \qquad (2.21)$$

holds. If p_i are the loop momenta, the integral associated with the Feynman diagram has the form

$$\mathcal{I}_G(k, m) = \int \prod_{i=1}^{L} \frac{\mathrm{d}^d p_i}{(2\pi)^d} \prod_{i=1}^{L} \frac{1}{(p_i + k_i)^2 + m^2} \prod_{j=1}^{V-1} \frac{1}{(\Delta p_j + k'_j)^2 + m^2}, \quad (2.22)$$

where k and k' are linear combinations of external momenta, with coefficients ± 1. Moreover, the Δp_j's are nontrivial linear combinations of the integrated momenta p with coefficients ± 1. We have used (2.18) to organize the integrand in the way shown.

We need to check the convergence of the integral in every region of integration. Since we are working in the Euclidean framework, the integral is regular for finite values of the momenta p. We just need to study its behavior when the momenta tend to infinity in all possible ways. It is sufficient to consider the following situations: i) the momenta of all the internal legs tend to infinity with the same velocity, or ii) the momenta of some internal legs are kept fixed, and the other ones tend to infinity with the same velocity. A singularity that occurs in case i) is called ultraviolet overall divergence. A singularity that occurs in case ii) is called ultraviolet subdivergence. Since in this book we treat only ultraviolet divergences, we omit to specify it from now on.

The overall divergences are studied by rescaling the integrated momenta p by a factor ξ,

$$p_i \to \xi p_i, \qquad (2.23)$$

and then sending ξ to infinity. The subdivergences are studied by performing the rescaling (2.23) with the constraint that the momenta of some internal legs are kept fixed. It can be shown that, once the divergences due to the two types of limits i) and ii) are cured, the integral becomes convergent. In other words, all the other ways to send the momenta to infinity are then automatically cured, because they amount to some combinations of the limits i) and ii). For example, if some momenta p_i' are rescaled by a factor ξ, and the other momenta p_i'' are rescaled by a factor ξ^2, then sending ξ to infinity is like first rescaling p_i'' by ξ at fixed p_i', then rescaling both p_i' and p_i'' by ξ again.

The subdivergences are due to the overall divergences of the irreducible subdiagrams G_{sub} of G. Precisely, G_{sub} can stand for any irreducible part of the subdiagram obtained by cutting the G internal legs whose momenta are kept fixed. Clearly, if G is irreducible, as we are going to assume from now on, the subdiagrams G_{sub} have fewer loops, because when we cut one or more G internal lines, we necessarily break some loop. Moreover, since the perturbative expansion, namely, the expansion in powers of $\hbar$, coincides with the loop expansion, the divergences can be subtracted algorithmically. In other words, when we deal with an L-loop diagram, we can assume it is already equipped with the set of counterterms that take care of its subdiagrams G_{sub}. For the moment, we ignore the subdivergences and concentrate on the overall divergences.

Let us compute the dimension of $\mathcal{I}_G(k, m)$. The momentum integration measure $\mathrm{d}^d p$ has dimension d, while the propagators have dimension -2. Using (2.18) and (2.21), we have

$$[\mathcal{I}_G(k, m)] = Ld - 2I = V \left[N \left(\frac{d}{2} - 1 \right) - d \right] - E \left(\frac{d}{2} - 1 \right) + d. \quad (2.24)$$

Make the rescaling (2.23) and consider the behavior of the integral when ξ tends to infinity. Let $\omega(\mathcal{I}_G)$ denote the power of ξ in this limit, called *degree of divergence* of the diagram G. In our present case, given the form of the integral (2.22), we have $\omega(\mathcal{I}_G) = [\mathcal{I}_G]$. However, if some external momentum or a mass factorizes, we may have $\omega(\mathcal{I}_G) < [\mathcal{I}_G]$. In general, we have the inequality $\omega(\mathcal{I}_G) \leqslant [\mathcal{I}_G]$. If $\omega(\mathcal{I}_G) < 0$ and there are no subdivergences, then the integral is ultraviolet convergent, because it is convergent in all the regions of integration. Instead, if $\omega(\mathcal{I}_G) \geqslant 0$, or $\omega(\mathcal{I}_G) < 0$, but there are subdivergences, the diagram is potentially ultraviolet divergent.

To begin with, consider a one-loop irreducible diagram. Since it has no 1PI subdiagrams, there can be only an overall divergence, but no subdivergences. Differentiate the diagram one time with respect to an external momentum k or a mass m, and observe that

$$\frac{\partial}{\partial k_\mu} \frac{1}{(p+k)^2+m^2} = -\frac{2(p+k)_\mu}{[(p+k)^2+m^2]^2},$$

$$\frac{\partial}{\partial m} \frac{1}{(p+k)^2+m^2} = -\frac{2m}{[(p+k)^2+m^2]^2}.$$

The differentiated diagram has a smaller degree of divergence:

$$\omega\left(\frac{\partial \mathcal{I}_G}{\partial K}\right) \leqslant \left[\frac{\partial \mathcal{I}_G}{\partial K}\right] = [\mathcal{I}_G] - 1,$$

where K is k_μ or m. Repeating the argument, we obtain

$$\omega\left(\frac{\partial^{n+r} \mathcal{I}_G}{\partial k_{\mu_1} \cdots \partial k_{\mu_n} \partial m^r}\right) \leqslant \left[\frac{\partial^{n+r} \mathcal{I}_G}{\partial k_{\mu_1} \cdots \partial k_{\mu_n} \partial m^r}\right] = [\mathcal{I}_G] - n - r.$$

If $n+r$ is sufficiently large, $[\mathcal{I}_G] - n - r$ becomes negative. Thus, if we differentiate the integral a sufficient number of times with respect to its external momenta and/or the masses, the integral becomes overall convergent. Said in equivalent words, the differentiation kills the overall divergent part. When we integrate back the result, we discover that the divergent part must be a polynomial of the masses and the external momenta. This is the crucial property of renormalization, and is called *locality* of the counterterms, because the Fourier transform of a polynomial of the momenta is a finite sum of delta functions and derivatives of delta functions, which are distributions localized at a single point.

We are still working on one-loop diagrams, but later on we show that the locality of the counterterms extends to the diagrams with arbitrarily many loops, provided we first subtract the subdivergences.

Now we describe how to subtract the overall divergent part of a one-loop diagram. Let $f(p,k,m)$ denote the unsubtracted integrand, and consider

$$\mathcal{I}_{G_R}(k,m) = \int \frac{d^d p}{(2\pi)^d}\left(f(p,k,m) - \sum_{n=0}^{\bar{\omega}} \frac{1}{n!} k_{\mu_1} \cdots k_{\mu_n} \frac{\partial_0^n f(p,k,m)}{\partial k_{\mu_1} \cdots \partial k_{\mu_n}}\right),$$

$$(2.25)$$

where $\bar{\omega}$ has to be determined, and the subscript 0 in ∂_0^n means that after taking the n derivatives with respect to k, k is set to zero. The sum in (2.25) collects the "counterterms". They remove the divergences from the integral.

In practice, we subtract the first $\bar{\omega}$ terms of the Taylor expansion of the integrand around vanishing external momenta. The integrand of $\mathcal{I}_{G_R}$ is still a rational function, and it is proportional to $\bar{\omega} + 1$ powers of the external momenta. Thus, we can write

$$\mathcal{I}_{G_R}(k, m) = k_{\mu_1} \cdots k_{\mu_{\bar{\omega}+1}} \int \frac{\mathrm{d}^d p}{(2\pi)^d} f_{\mu_1 \cdots \mu_{\bar{\omega}+1}}(p, k, m),$$

for some other rational functions $f_{\mu_1 \cdots \mu_{\bar{\omega}+1}}$. Now,

$$\omega(\mathcal{I}_{G_R}) \leqslant [\mathcal{I}_{G_R}] - \bar{\omega} - 1 = [\mathcal{I}_G] - \bar{\omega} - 1.$$

If we choose $\bar{\omega} = [\mathcal{I}_G]$, we obtain $\omega(\mathcal{I}_{G_R}) < 0$, which means that $\mathcal{I}_{G_R}$ is overall convergent.

For example, consider the one-loop correction to the four-point function of the theory φ_4^4. We have the sum of the three diagrams

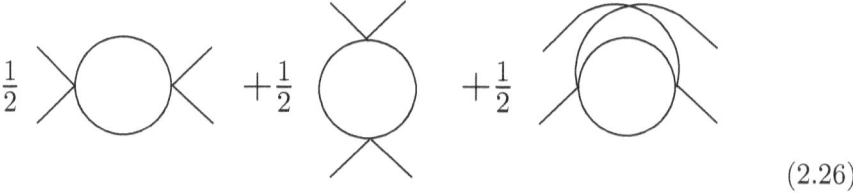

$$\frac{1}{2} \qquad + \frac{1}{2} \qquad + \frac{1}{2} \qquad \qquad \qquad (2.26)$$

each of which has the form $(\lambda^2/2)\mathcal{I}(k, m)$, where

$$\mathcal{I}(k, m) = \int \frac{\mathrm{d}^4 p}{(2\pi)^4} \frac{1}{p^2 + m^2} \frac{1}{(p + k)^2 + m^2}, \qquad (2.27)$$

with different combinations k of the external momenta. The integral (2.27) has $\omega = 0$ and a logarithmic divergence. The subtracted integral reads

$$\mathcal{I}_R(k, m) = -k_\mu \int \frac{\mathrm{d}^4 p}{(2\pi)^4} \frac{2p_\mu + k_\mu}{(p^2 + m^2)^2((p + k)^2 + m^2)},$$

which is clearly convergent.

We have successfully subtracted the one-loop integrals. The question is whether the procedure makes physical sense, or we have arbitrarily changed the theory we started from. Here enters the crucial property of counterterms, which is their locality. Formula (2.25) shows that, working in momentum

space, the counterterms are polynomials of the external momenta. For example, the counterterm for (2.27) is

$$R(k, m) \equiv \mathcal{I}_R(k, m) - \mathcal{I}(k, m) = -\int \frac{d^4p}{(2\pi)^4} \frac{1}{(p^2 + m^2)^2}, \qquad (2.28)$$

which is k independent, i.e., just a (divergent) constant.

While a diagram is a nonlocal function of the external momenta, its
divergent part is local. Thanks to this property, it looks like a vertex, or an
inverse propagator. For this reason, it can be subtracted by adding *ad hoc*
local terms to the action. To do this efficiently, however, we have to use a
specific regularization technique, because the integral (2.28) is meaningless
without it. An explicit regularization, like the dimensional one, allows us to
separate $\mathcal{I}$ from $\mathcal{R}$, and move the counterterms around at will. We stress that
the use of a regularization is not necessary to define perturbative quantum
field theory, and some regularization independent approaches are advertised
in the literature. However, we do not recommend the reader to renounce the
practical convenience of working with an explicit regularization, in order to
keep track of what is going on. It is relatively easy to prove that the physical
results are regularization independent anyway (see below).

Let us switch from the cutoff regularization to the dimensional regularization. The integral of (2.27) is promoted to D dimensions as

$$\mathcal{I}_D(k, m) = \int \frac{d^D p}{(2\pi)^D} \frac{1}{p^2 + m^2} \frac{1}{(p + k)^2 + m^2}. \qquad (2.29)$$

and each λ gets multiplied by μ^ε. Using Feynman parameters, namely,
formula (A.2), we can rewrite the integral as

$$\int_0^1 dx \int \frac{d^D p}{(2\pi)^D} \frac{1}{((p + kx)^2 + m^2 + k^2 x(1 - x))^2}. \qquad (2.30)$$

Then, we can make a translation $p \to p - kx$ and use (A.4). We get

$$\frac{\lambda^2 \mu^{2\varepsilon}}{2} \mathcal{I}_D(k, m) = \frac{\lambda^2 \mu^{2\varepsilon} \Gamma\left(2 - \frac{D}{2}\right)}{2(4\pi)^{D/2}} \int_0^1 dx \left(k^2 x(1 - x) + m^2\right)^{D/2 - 2}$$

$$= \frac{\lambda^2 \mu^\varepsilon}{16\pi^2 \varepsilon} + \frac{\lambda^2 \mu^\varepsilon}{32\pi^2} \left(2 - \gamma_E + \ln \frac{4\pi \mu^2}{m^2}\right) \qquad (2.31)$$

$$-2\sqrt{1 + \frac{4m^2}{k^2}} \operatorname{arcsinh} \sqrt{\frac{k^2}{4m^2}}\right) + \mathcal{O}(\varepsilon).$$

The counterterm (2.28) becomes

$$-\frac{\lambda^2 \mu^{2\varepsilon}}{2} \int \frac{d^D p}{(2\pi)^D} \frac{1}{(p^2 + m^2)^2} = -\frac{\lambda^2}{2} \frac{\Gamma\left(2 - \frac{D}{2}\right)}{(4\pi)^{D/2}} \mu^\varepsilon \left(\frac{\mu}{m}\right)^\varepsilon$$

$$= -\lambda^2 \mu^\varepsilon \left(\frac{1}{16\pi^2 \varepsilon} + c_1\right), \qquad (2.32)$$

where the constant c_1 is finite in the limit $\varepsilon \to 0$. Since the diagrams of (2.26) are three, and all of them have the same divergent part, we have to modify the action so as to subtract three times the divergent part of (2.32). This result can be achieved by adding

$$\Delta\mathcal{L} = 3\lambda^2 \mu^\varepsilon \left(\frac{1}{16\pi^2 \varepsilon} + c_1\right) \frac{\varphi^4}{4!} \qquad (2.33)$$

to the Lagrangian. Note that the power $\mu^{2\varepsilon}$ provided by the diagram, shown in the first line of (2.31), has become μ^ε in the counterterm (2.33), to match the dimensions correctly. The other factor μ^ε gets expanded in ε. As a consequence, μ can enter the logarithms in the right places to make their arguments dimensionless, such as in the second line of (2.31).

Now we note that the constant c_1 appearing in (2.33) does not actually need to be the one of (2.32), because what is important is to subtract the divergent part. Thus, the c_1 of formula (2.33) can be arbitrary. Roughly speaking, when we subtract infinity, we may as well subtract infinity plus any finite constant we want. Later on we will see that the physical quantities do not depend on this arbitrariness.

The correction (2.33) produces an extra vertex

$$= -3\lambda^2 \mu^\varepsilon \left(\frac{1}{16\pi^2 \varepsilon} + c_1\right)$$

$$(2.34)$$

which must be added to the Feynman rules. The vertex (2.34) carries an additional (hidden) power of $\hbar$, since it is of order λ^2. Diagrammatically, it counts like a one-loop diagram, so it appears in the right place to subtract the divergences of (2.26). The finite value of a single subtracted diagram of (2.26) is thus

$$-\frac{\lambda^2 \mu^\varepsilon}{16\pi^2} \left(\sqrt{1 + \frac{4m^2}{k^2}} \, \text{arcsinh}\sqrt{\frac{k^2}{4m^2}} - \frac{1}{2} \ln \frac{4\pi \mu^2}{m^2} + c\right), \qquad (2.35)$$

where c is an arbitrary finite, k-independent constant. We may as well assume that c is independent of μ and m. The result admits a smooth massless limit

$$\frac{\lambda^2 \mu^\varepsilon}{32\pi^2} \left(\ln \frac{4\pi\mu^2}{k^2} - 2c \right), \tag{2.36}$$

which can also be easily computed from (2.29). Indeed, using (A.3) to calculate the integral over x, the first line of (2.31) gives, at $m = 0$,

$$\frac{\lambda^2 \mu^{2\varepsilon}}{2} \mathcal{I}_D(k, 0) = \frac{\lambda^2 \mu^{2\varepsilon} \Gamma\left(2 - \frac{D}{2}\right) \left[\Gamma\left(\frac{D}{2} - 1\right)\right]^2}{2(4\pi)^{D/2} \Gamma(D - 2)} \left(k^2\right)^{D/2 - 2}. \tag{2.37}$$

Note again that the factor in front of expression (2.36) is μ^ε instead of $\mu^{2\varepsilon}$, and that the argument of the logarithm contains appropriate factors of μ, which make it dimensionless.

The modification (2.33), which subtracts the divergence away, does not look so serious after all. In the end, it just amounts to redefining the coupling constant in front of φ^4. We are certainly allowed to do that, since we have not attached any physical meaning to λ, so far. This is the idea of *renormalization*, and justifies its name. It is the removal of the divergences by means of redefinitions of fields and parameters. Note that it would not be possible to achieve this goal if the counterterms were not local, since the action is local by assumption. At the same time, locality alone is not sufficient to ensure that the divergences can be renormalized.

Consider for example the theory φ_4^6. We write the interacting Lagrangian as

$$\mathcal{L}_I = \lambda_6 \mu^{2\varepsilon} \frac{\varphi^6}{6!},$$

where λ_6 is a coupling constant of dimension -2. At one loop we have divergent diagrams such as

$$\tag{2.38}$$

The corresponding integral is again proportional to $\mathcal{I}_D(k, m)$. However, to subtract this kind of divergence we need to modify the Lagrangian by adding a counterterm of the form

$$\Delta\mathcal{L} = 35\lambda_6^2 \mu^{3\varepsilon} \left(\frac{1}{16\pi^2\varepsilon} + c_1 \right) \frac{\varphi^8}{8!}, \tag{2.39}$$

where 35= 8!/(2!4!4!) is the number of nontrivial permutations of the external legs. The modified action contains an interaction, φ^8, that is not present in the action of the theory φ_4^6. Therefore, (2.39) cannot be absorbed into a simple redefinition of the fields and the couplings, but demands a radical modification of the theory, from a φ^6 potential to a $\varphi^6 + \varphi^8$ potential. Moreover, that modification is not even sufficient. Using two vertices φ^8, we can easily construct a one-loop diagram similar to (2.38), with 6+6 external legs. Again, it is logarithmically divergent, and its divergent part can be subtracted only at the price of introducing a φ^{12} vertex. We can go on like this indefinitely: we discover that the renormalization of divergences is possible only at the price of introducing infinitely many new vertices and new independent couplings.

Concluding, the locality of counterterms is necessary, but not sufficient, to have control on the divergences. We need to check that all the counterterms have the form of the terms that are already contained in the initial Lagrangian. When that happens, the divergences can be removed by redefining the fields and the couplings, the subtraction of divergences is a stable procedure and the final Lagrangian is a simple redefinition of the initial one. Otherwise, we can attempt to stabilize the Lagrangian, by adding new *ad hoc* local terms. Next, we must check that a finite number of such new terms is sufficient to stabilize the subtraction of divergences to all orders. If that does not happen, the final Lagrangian contains infinitely many independent couplings and interactions.

The theories that contain finitely many vertices and are stable under renormalization are called renormalizable. The theories that are not stable under renormalization, because they end up containing infinitely many independent terms, are called nonrenormalizable. As we will prove, the theory φ_4^4 is renormalizable. We have already proved that the theory φ_4^6 cannot be stabilized, so it is nonrenormalizable.

The nonrenormalizable theories are described by nonpolynomial Lagrangians, which are the sums of local terms with arbitrarily high powers of the fields and their derivatives. We have

$$\mathcal{L}_{\text{nonren}} = \frac{1}{2}(\partial_\mu \varphi)^2 + \frac{m^2}{2}\varphi^2 + \sum_{\{m,n\}} \frac{\lambda_{\{m,n\}}}{M^{X(m,n)}} \prod_i (\partial^{m_i} \varphi^{n_i}), \qquad (2.40)$$

where M is some energy scale and

$$X(m, n) = \sum_i \left(m_i + n_i \frac{d-2}{2} \right) - d$$

is chosen to make the couplings $\lambda_{\{m,n\}}$ dimensionless. The nonrenormaliz-able theories are problematic from the physical point of view. Their corre-lation functions depend on infinitely many parameters, which means, at the practical level, that infinitely many measurements are necessary to determine the theory completely, and make predictions that are valid at arbitrarily high energies.

On the other hand, in most cases a nonrenormalizable theory can be used to make predictions at low energies. If a monomial $\mathcal{O}$ of the fields and their derivatives has dimension $d_{\mathcal{O}}$, then its insertion into a correlation function behaves like $E^{d_{\mathcal{O}}}$ at low energies, so the interacting Lagrangian behaves like

$$\mathcal{L}_I \sim E^d \sum_{\{m,n\}} \lambda_{\{m,n\}} \left(\frac{E}{M} \right)^{X(m,n)}.$$

We can have three typical cases.

1) If all the dimensionless couplings $\lambda_{\{m,n\}}$ are of comparable orders, only a finite number of interactions is important at energies $E \ll M$. We then say that *almost all* the interactions become negligible at low energies. However, the number of interactions that are important grows with the energy and becomes infinite at $E \sim M$.

2) A behavior like $\lambda_{\{m,n\}} \sim X(m, n)^{-X(m,n)}$ for large m, n, ensures that almost all the couplings can be neglected in every energy range bounded from above. The number of important couplings grows with the energy and becomes infinite only at $E = \infty$.

3) A behavior like $\lambda_{\{m,n\}} \sim X(m, n)^{X(m,n)}$ for large m, n, implies that the interaction multiplied by the parameter $\lambda_{\{m,n\}}$ is negligible for energies

$$E \ll \frac{M}{X(m, n)}.$$

Then there exists no energy range where almost all the couplings can be neglected.

Intermediate types of behaviors can be traced back to these three cases. The behaviors of the couplings $\lambda_{\{m,n\}}$ are a priori unknown, but the comparison with experiments can to some extent suggest whether we are in the situations 1), 2) or 3).

Even in the worst case, a nonrenormalizable theory may still have a nontrivial predictive content. Indeed, even if the Lagrangian contains infinitely many independent unknown parameters, there might still exist physical quantities that just depend on a finite subset of them. The hard part is to work out those physical quantities and make experiments that are suitable for them. Strictly speaking, the difference between renormalizable and nonrenormalizable theories is that the former are always predictive, in an obvious way, while the latter can be predictive, but only in a rather nontrivial way.

It is worth to stress that the nonrenormalizable theories are much less problematic from the mathematical point of view, where it does not really matter whether the number of independent couplings is finite or infinite. Indeed, most renormalization theorems we are going to prove hold both for renormalizable and nonrenormalizable theories.

φ_4^4 at one loop

Let us complete the one-loop renormalization of the φ_4^4 theory. Formula (2.24) gives

$$\omega_G \leqslant 4 - E,$$

so the potentially divergent diagrams, which are those with $\omega_G \geqslant 0$, have $E \leqslant 4$. The renormalization of the four-point function has been discussed above. Since the φ_4^4 theory has a $\mathbb{Z}_2$ symmetry $\varphi \to -\varphi$, the correlation functions that contain an odd number of insertions are identically zero. Moreover, the diagrams with zero external legs need not be considered, since they can always be subtracted by adding a constant to the Lagrangian. We remain with the one-loop correction to the two-point function, which is the second term on the right-hand side of (1.46), and gives the integral

$$-\frac{\lambda \mu^\varepsilon}{2} \int \frac{\mathrm{d}^D p}{(2\pi)^D} \frac{1}{p^2 + m^2} = \lambda m^2 \left(\frac{1}{16\pi^2 \varepsilon} - c_2 \right), \tag{2.41}$$

where the constant c_2 is regular for $\varepsilon \to 0$. To subtract this divergence we modify the action by adding

$$\Delta' \mathcal{L} = \lambda m^2 \left(\frac{1}{16\pi^2 \varepsilon} - c_2 \right) \frac{\varphi^2}{2}. \tag{2.42}$$

Again, we can choose an arbitrary finite c_2 here, different from the one appearing in (2.41). Collecting (2.33) and (2.42), the full one-loop renormalized action reads

$$S_1(\varphi) = \int d^D x \left(\frac{1}{2}(\partial_\mu \varphi)^2 + m^2 \left(1 + \frac{\lambda}{16\pi^2 \varepsilon} - \lambda c_2\right) \frac{\varphi^2}{2} \right.$$
$$\left. + \lambda \mu^\varepsilon \left(1 + \frac{3\lambda}{16\pi^2 \varepsilon} + 3\lambda c_1\right) \frac{\varphi^4}{4!} \right). \qquad (2.43)$$

More generally, the renormalized action can be written as

$$S_R(\varphi, \lambda, m, \mu) = \int d^D x \left(\frac{Z_\varphi}{2}(\partial_\mu \varphi)^2 + m^2 Z_{m^2} \frac{Z_\varphi \varphi^2}{2} + \lambda \mu^\varepsilon Z_\lambda \frac{Z_\varphi^2 \varphi^4}{4!} \right),$$
$$(2.44)$$

where the coefficients

$$Z_\varphi = 1 + \mathcal{O}(\lambda^2), \qquad Z_{m^2} = 1 + \frac{\lambda}{16\pi^2 \varepsilon} - \lambda c_2 + \mathcal{O}(\lambda^2),$$
$$Z_\lambda = 1 + \frac{3\lambda}{16\pi^2 \varepsilon} + 3\lambda c_1 + \mathcal{O}(\lambda^2), \qquad (2.45)$$

which depend on λ and ε, are called "renormalization constants". At one loop, S_R coincides with S_1. If we collect the field and parameter redefinitions into the "bare" quantities

$$\varphi_B = Z_\varphi^{1/2} \varphi, \qquad m_B^2 = m^2 Z_{m^2}, \qquad \lambda_B = \lambda \mu^\varepsilon Z_\lambda, \qquad (2.46)$$

then $S_R(\varphi, \lambda, m, \mu)$ can be rewritten in bare form

$$S_B(\varphi_B, \lambda_B, m_B) \equiv \int d^D x \left(\frac{1}{2}(\partial_\mu \varphi_B)^2 + m_B^2 \frac{\varphi_B^2}{2} + \lambda_B \frac{\varphi_B^4}{4!} \right). \qquad (2.47)$$

We see that the bare action is exactly the classical action.

We have already observed that the constants c_1 and c_2 of formula (2.43) are arbitrary. Any time we subtract a pole $1/\varepsilon$, we can equivalently subtract $1/\varepsilon$ plus a finite constant. This arbitrariness amounts to a finite redefinition of the fields and the parameters, which has no physical significance.

A specific prescription to choose such arbitrary constants is called subtraction scheme. For example, subtracting the first terms of the Taylor expansion around vanishing external momenta is a scheme prescription. In massless theories this prescription is not convenient, because it can originate spurious infrared divergences. In that case it is better, for example, to subtract the first terms of the Taylor expansion around some nontrivial configurations of the external momenta. We can even choose different configurations for different diagrams.

A very popular scheme, called *minimal* subtraction scheme, amounts to subtracting just the poles in ε, with no finite parts attached. Variants of the minimal subtraction scheme amount to subtracting the poles in ε plus certain universal finite parts.

The constants c_1 and c_2 parametrize the scheme arbitrariness at one loop. The residues of the poles $1/\varepsilon$, on the other hand, are scheme independent. For example, comparing (2.3) and (2.7), we have remarked that the coefficients of $\ln \Lambda$ and $1/\varepsilon$ coincide, as well as the term $m^2 \ln m^2$. Instead, the quadratic divergences Λ^2 end into the arbitrary constant c_2.

A few tricks can allow us to compute the divergent parts quite easily, taking advantage of their locality. Consider for example the integral $\mathcal{I}_D(k, m)$ of formula (2.29). We know that its divergent part is a polynomial of k and m of degree zero. Therefore, it is just a constant, and can be calculated by setting k and m to the values we like. We cannot put both $k = m = 0$, however, because this affects the domain of integration in a nontrivial way. The rules of the dimensional regularization do not allow us to exchange the integral and the limits $k \to 0$, $m \to 0$ in this case. A better option is to keep $m \neq 0$ and put $k = 0$: since the domain of integration is unaffected by this choice, the limit $k \to 0$ can be safely taken inside the integral. We could also keep $k \neq 0$ and put $m = 0$, but the first option is more convenient. Then, (2.29) becomes a standard integral (see Appendix A), and its divergent part can be worked out immediately.

More generally, since the divergent part of a diagram is a polynomial of the external momenta k and the masses m, if we differentiate the integral with respect to k and m, we can reduce the degree of the polynomial to zero, and then proceed as above. If we differentiate in all possible ways, we can fully reconstruct the polynomial, i.e., the divergent part of the diagram.

Using these tricks,

Exercise 4 *compute the one-loop renormalization of the φ_6^3 theory.*

Solution. The renormalized action reads

$$S(\varphi) = \int \mathrm{d}^D x \left(\frac{Z_\varphi}{2} (\partial_\mu \varphi)^2 + \frac{m^2}{2} Z_{m^2} Z_\varphi \varphi^2 + \lambda \mu^\varepsilon Z_\lambda Z_\varphi^{3/2} \frac{\varphi^3}{3!} + m^4 \mu^{-\varepsilon} \Delta_1 \varphi \right),$$

where $\varepsilon = 3 - D/2$. At one loop the divergent diagrams are those with one, two and three external legs. The tadpole (diagram with a single external leg) is

$$-\lambda\mu^\varepsilon \frac{\Gamma\left(1 - \frac{D}{2}\right) m^{D-2}}{(4\pi)^{D/2}} = -\frac{\lambda m^4 \mu^{-\varepsilon}}{2(4\pi)^3 \varepsilon} + \text{finite},$$

whence

$$\Delta_1 = -\frac{\lambda}{2(4\pi)^3 \varepsilon}.$$

The self-energy (diagram with two external legs) is equal to the first line of (2.31). The difference is that now we have to expand it around $D = 6$, instead of $D = 4$. We obtain the divergent part

$$-\frac{\lambda^2}{12\varepsilon(4\pi)^3}\left(k^2 + 6m^2\right),$$

which gives

$$Z_\varphi = 1 - \frac{\lambda^2}{12\varepsilon(4\pi)^3} + \mathcal{O}(\lambda^4), \qquad Z_{m^2} = 1 - \frac{5\lambda^2}{12\varepsilon(4\pi)^3} + \mathcal{O}(\lambda^4). \qquad (2.48)$$

The divergent part of the correction to the vertex can be calculated at vanishing external momenta. We have

$$-\int \frac{\mathrm{d}^D p}{(2\pi)^D} \frac{\lambda^3 \mu^{3\varepsilon}}{(p^2 + m^2)^3} = -\lambda^3 \mu^{3\varepsilon} \frac{\Gamma(3 - D/2)}{2(4\pi)^{D/2}} (m^2)^{D/2-3} = -\frac{\lambda^3 \mu^\varepsilon}{2\varepsilon(4\pi)^3} + \text{finite},$$

so the vertex renormalization constant is

$$Z_\lambda = 1 - \frac{3\lambda^2}{8\varepsilon(4\pi)^3} + \mathcal{O}(\lambda^4). \qquad (2.49)$$

□

Most properties of the one-loop renormalization generalize to all orders. Now we make some remarks about the renormalization at two loops, which help us introduce the proofs of all-order statements.

φ_4^4 at two loops

We denote the vertices provided by the one-loop counterterms (2.33) and (2.42) with a dot, as in (2.34). At two loops, we have diagrams that contain, in general, both subdivergences and overall divergences. For example, the two-loop corrections to the four-point function are given by the diagrams

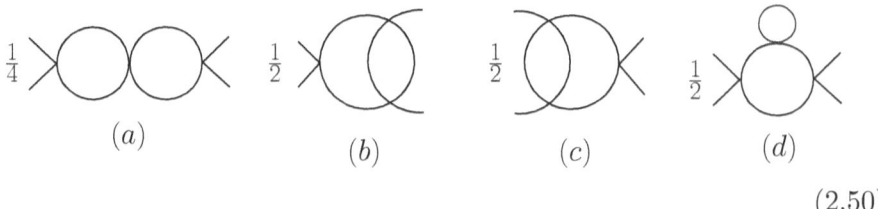

$$(2.50)$$

plus two permutations of each. We begin by concentrating on the first three diagrams, since the fourth one is much simpler to deal with. The subdivergences of the diagrams (a), (b) and (c) are subtracted by

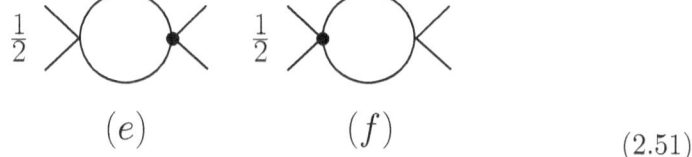

$$(2.51)$$

Precisely, the subdivergences of each subdiagram of (a), (b) and (c) are given by one third of (2.34). Moreover, (2.34) also absorbs a combinatorial factor $1/2$, which is the combinatorial factor of the diagrams (2.26). It is convenient to define separate counterterms for the three diagrams (2.26), even if they are equal in value. We do this by splitting the counterterm (2.34) into the sum of three equal contributions, and using appropriate labels to remember which diagram they cure. By doing so, we obtain

$$(2.52)$$

Observe that the rules to compute the combinatorial factors remain the same after this splitting. If A is a vertex with N external legs, and V is the number of times it appears in a diagram, its contribution is $A^V/(N!^V V!)$. Now, if A is decomposed as a sum $\sum_{i=1}^{n} a_i$, the multinomial formula ensures that each "subvertex" contributes according to the same rule. Indeed, we have

$$\frac{A^V}{(N!)^V V!} = \frac{(\sum_{i=1}^{n} a_i)^V}{(N!)^V V!} = \sum_{\{n_i\}} \prod_{i=1}^{n} \frac{a_i^{n_i}}{(N!)^{n_i} n_i!}, \qquad (2.53)$$

where the sum is over sets of nonnegative n_i's such that $\sum_{i=1}^{n} n_i = V$. Note that it is not necessary that each term a_i of the sum be symmetrized under the exchange of its external legs.

Now, consider the diagram (a) and its counterterms:

$$(2.54)$$

This sum is free of subdivergences. Indeed, the two subdiagrams of (a) are of the first type of the list that appears on the right-hand side of (2.52). Moreover, each subdiagram carries a combinatorial factor $1/2$, which explains why the counterterms in (2.54) are correctly multiplied by $1/2$ instead of $1/4$.

Next, consider the diagram (b): the sum

$$(2.55)$$

is also free of subdivergences. Observe that this time we use the second and third vertices of (2.52), because they both correspond to the divergent subdiagram of (b). Again the combinatorial factors match, taking into account the factor $1/2$ absorbed by the divergent subdiagram. The diagram (c) is treated symmetrically to (b).

In conclusion, the sums

$$s_1 = (a) + \frac{1}{3}(e) + \frac{1}{3}(f), \qquad s_2 = (b) + \frac{2}{3}(e), \qquad s_3 = (c) + \frac{2}{3}(f). \quad (2.56)$$

are all free of subdivergences. Therefore, so is the total $t = s_1 + s_2 + s_3 = (a) + (b) + (c) + (e) + (f)$.

Since s_1, s_2 and s_3 are free of subdivergences, so are their derivatives with respect to the external momenta and the masses. Now, a sufficient number of such derivatives kills the overall divergences of s_1, s_2 and s_3, and produces fully convergent integrals. This proves that the divergent parts of the subtracted integrals s_1, s_2 and s_3 are polynomial in the masses and the external momenta.

Let us explicitly check this fact in s_1. For simplicity, we work at $m = 0$. The diagram (a) is easy to calculate, since it is basically the square of any diagram of (2.26). We have

$$(a) = \frac{(-\lambda)^3 \mu^\varepsilon}{4} \left(\mu^\varepsilon \mathcal{I}_D(k, 0) \right)^2 .$$

Write

$$\mu^\varepsilon \mathcal{I}_D(k, 0) = \frac{a}{\varepsilon} + b \ln \frac{k^2}{\mu^2} + c,$$

where a, b and c are finite for $\varepsilon \to 0$. Their values can be read from the calculations already made (in particular, $a = 1/(8\pi^2)$), but for what we are going to say we do not even need them. Each counterterm of (2.54) equals $-a\lambda^2 \mu^\varepsilon/(2\varepsilon)$, so

$$\frac{1}{3}(e) = \frac{1}{3}(f) = -\frac{1}{2\varepsilon} a\lambda^2 \mu^\varepsilon \left(\frac{-\lambda \mu^\varepsilon}{2} \right) \mathcal{I}_D(k, 0).$$

Finally,

$$s_1 = -\frac{\lambda^3 \mu^\varepsilon}{4} \left[\left(\frac{a}{\varepsilon} + b \ln \frac{k^2}{\mu^2} + c \right)^2 - 2\frac{a}{\varepsilon} \left(\frac{a}{\varepsilon} + b \ln \frac{k^2}{\mu^2} + c \right) \right].$$

In the difference, the (nonlocal) subdivergences

$$2\frac{ab}{\varepsilon} \ln \frac{k^2}{\mu^2} \tag{2.57}$$

subtract away, and the surviving (overall) divergences are purely local. We find

$$s_1 = -\frac{\lambda^3 \mu^\varepsilon}{4} \left(-\frac{a^2}{\varepsilon^2} + \text{finite part} \right), \tag{2.58}$$

as expected.

This example, although very simple, is sufficient to illustrate the most general facts. The subdivergences are in general nonlocal, because they are "products" of a divergent part, originated by some subdiagram, times a finite (thus nonlocal) part, due to the rest of the diagram. Subtracting something like (2.57) would require to alter the original theory completely, turning it into a nonlocal theory. Fortunately, the subdivergences are automatically subtracted by the counterterms associated with the subdiagrams.

It remains to consider the diagram (d) of (2.50). Its subdivergence is subtracted by

$$(g) \tag{2.59}$$

where the dot denotes the counterterm of (2.42). The sum $s_4 = (d) + (g)$ is clearly free of subdivergences.

Consider now the two-loop corrections to the two-point function, which are given by the diagrams

$$\frac{1}{4} \quad \underbrace{\qquad}_{(h)} \qquad \frac{1}{3!} \quad \underbrace{\qquad}_{(k)} \tag{2.60}$$

The counterterms that subtract the subdivergences are

$$\frac{1}{2} \quad \underbrace{\qquad}_{(i)} \qquad \frac{1}{2} \quad \underbrace{\qquad}_{(j)} \tag{2.61}$$

which vanish at vanishing masses, since they are tadpoles. The diagram (h) has two types of diverging subdiagrams, corresponding to both types of contributions (i) and (j). Instead, the diagram (k) has a single type of subdiagram, but it appears three times, since freezing any internal line gives the same result. This turns the combinatorial factor $1/3!$ into $1/2$. The total $t' = (h) + (k) + (i) + (j)$ can be arranged as $s_5 + s_6$, where

$$s_5 = (h) + \frac{1}{3}(i) + (j), \qquad s_6 = (k) + \frac{2}{3}(i), \tag{2.62}$$

are both free of subdivergences. Explicitly, the (h) subdivergence due to the bottom loop is subtracted by $(i)/3$, while the (h) subdivergence due to the top loop is subtracted by (j). This $(i)/3$ is obtained by using the middle vertex of (2.52). Similarly, the (k) subdivergences are subtracted by the diagrams of type (i) obtained from the first and third vertices of (2.52). You can check this fact by working directly on the vertices of (2.52) (attaching the external lines and connecting the internal lines), and recalling that they match the diagrams of (2.26). Note that the factor 2 of $2(i)/3$ compensates for the combinatorial factor in front of (i), and returns the right coefficient.

Again, this proves that the overall divergences of the sums s_5 and s_6 are polynomials of the masses and the external momenta.

Exercise 5 *Calculate Z_φ at two loops in the φ_4^4 theory.*

Solution. Since Z_φ is independent of m, it is the same in the massless and massive theories, so we can work at $m = 0$. The two-loop contribution to the self-energy is

$$\frac{\lambda^2 \mu^{2\varepsilon}}{3!} \int \frac{d^D p}{(2\pi)^D} \frac{d^D q}{(2\pi)^D} \frac{1}{p^2 q^2 (p + q + k)^2}, \tag{2.63}$$

where k is the external momentum. We first integrate on p by means of formula (2.37), which gives

$$\frac{\lambda^2 \mu^{2\varepsilon}}{3!} \frac{\Gamma\left(2 - \frac{D}{2}\right) \left[\Gamma\left(\frac{D}{2} - 1\right)\right]^2}{(4\pi)^{D/2} \Gamma(D - 2)} \int \frac{d^D q}{(2\pi)^D} \frac{1}{q^2 [(q + k)^2]^{2 - D/2}}.$$

Then we use the Feynman parameters again to calculate the integral over q. We obtain

$$\frac{\lambda^2 \mu^{2\varepsilon}}{3!} \frac{\Gamma(3 - D) \left[\Gamma\left(\frac{D}{2} - 1\right)\right]^3}{(4\pi)^D \Gamma\left(\frac{3D}{2} - 3\right)} (k^2)^{D-3}. \tag{2.64}$$

Extracting the divergent part, we finally find

$$Z_\varphi = 1 - \frac{\lambda^2}{12\varepsilon (4\pi)^4} + \mathcal{O}(\lambda^3). \tag{2.65}$$

Exercise 6 *Calculate Z_λ at two loops in the φ_4^4 theory.*

Solution. The diagrams we have to study are those of formula (2.50), plus the counterterms (2.51) and (2.59), plus two permutations of each. Since Z_λ is independent of m, it is the same in the massless and massive theories. The diagram (d) and its counterterm (g) can be ignored, since they vanish at $m = 0$.

Besides being independent of m, Z_λ is independent of the external momenta k. We can simplify the calculation by setting k to zero and working with nonvanishing artificial masses δ, to avoid infrared problems. Alternatively, we can set the masses equal to zero and choose convenient configurations of the external momenta. We adopt the second option. The divergent contributions s_3^{div} and s_2^{div} of s_3 and s_2 coincide, so the total divergent part can be written as $3(s_1^{\mathrm{div}} + 2s_2^{\mathrm{div}})$, where s_1^{div} can be read from (2.58) and the overall factor 3 takes the permutations of the external legs into account.

Now we evaluate the diagram (b). Let k denote the total incoming momentum of the two external legs on the left-hand side. We can simplify the calculation by setting the momentum k' of the top-right external leg to zero. Indeed, it is easy to see that the integral becomes fully convergent after one derivative with respect to k', which means that the divergent part, although nonlocal, cannot depend on k'. The same trick does not work for k, so we cannot set $k = 0$. The subdiagram can be replaced with its exact expression (2.37). We get

$$(b) = -\frac{\lambda^3 \mu^{3\varepsilon} \Gamma\left(2 - \frac{D}{2}\right) \left[\Gamma\left(\frac{D}{2} - 1\right)\right]^2}{2(4\pi)^{D/2} \Gamma(D-2)} \int \frac{\mathrm{d}^D p}{(2\pi)^D} \frac{1}{(p^2)^{3-D/2}(p-k)^2}.$$

Now the calculation proceeds as usual. We find

$$(b) = \frac{\lambda^3 \mu^{3\varepsilon} \left[\Gamma\left(\frac{D}{2} - 1\right)\right]^3 \Gamma(3 - D)}{(4\pi)^D (4 - D)\Gamma\left(\frac{3D}{2} - 4\right)} (k^2)^{D-4}$$

$$= -\frac{\lambda^3 \mu^{\varepsilon}}{(4\pi)^4 \varepsilon^2} - \frac{\lambda^3 \mu^{\varepsilon}}{(4\pi)^4 \varepsilon} \left(\frac{5}{2} - \gamma_E - \ln\frac{k^2}{(4\pi)\mu^2}\right) + \text{finite}.$$

On the other hand, it is easy to evaluate (e), which gives

$$(e) = \frac{3\lambda^3 \mu^{2\varepsilon} \Gamma\left(2 - \frac{D}{2}\right) \left[\Gamma\left(\frac{D}{2} - 1\right)\right]^2}{2(4\pi)^2 \varepsilon (4\pi)^{D/2} \Gamma(D-2)} (k^2)^{D/2-2}$$

$$= \frac{3\lambda^3 \mu^{\varepsilon}}{(4\pi)^4 \varepsilon^2} + \frac{3\lambda^3 \mu^{\varepsilon}}{2(4\pi)^4 \varepsilon} \left(2 - \gamma_E - \ln\frac{k^2}{(4\pi)\mu^2}\right) + \text{finite}.$$

The total is

$$s_2 = (b) + \frac{2}{3}(e) = \frac{\lambda^3 \mu^{\varepsilon}}{(4\pi)^4 \varepsilon} \left(\frac{1}{\varepsilon} - \frac{1}{2}\right) + \text{finite}.$$

Note that the nonlocal subdivergences did cancel out, as expected. Finally, collecting the contributions of s_1 and s_2, we get

$$3(s_1^{\mathrm{div}} + 2s_2^{\mathrm{div}}) = \frac{3\lambda^3 \mu^{\varepsilon}}{(4\pi)^4 \varepsilon} \left(\frac{3}{\varepsilon} - 1\right).$$

Using (2.65) we obtain

$$Z_\lambda(\lambda, \varepsilon) = 1 + \frac{3\lambda}{(4\pi)^2\varepsilon} + \frac{9\lambda^2}{(4\pi)^4\varepsilon^2} - \frac{17\lambda^2}{6(4\pi)^4\varepsilon} + \mathcal{O}(\lambda^3). \tag{2.66}$$

Exercise 7 *Compute the two-loop renormalization of the massless φ_3^6 theory.*

Solution. The renormalized action is

$$S(\varphi) = \int \mathrm{d}^D x \left(\frac{Z_\varphi}{2}(\partial_\mu \varphi)^2 + \lambda \mu^{2\varepsilon} Z_\lambda Z_\varphi^3 \frac{\varphi^6}{6!} \right),$$

where $\varepsilon = 3 - D$. It is easy to check that there is no one-loop divergence, so we just have to consider the two-loop diagrams. Moreover, there is no two-loop contribution to the wave-function renormalization constant. Instead, the vertex gets a counterterm from the diagram

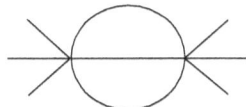

The combinatorial factor is $1/6$. The divergent part does not depend on the permutations of the external legs, which gives an extra factor 10. In the end, we just have to calculate (2.63), and multiply it by $10\mu^{2\varepsilon}$. Using (2.64) and expanding around $D = 3$ we obtain

$$Z_\lambda = 1 + \frac{5\lambda}{6\varepsilon(4\pi)^2} + \mathcal{O}(\lambda^2). \tag{2.67}$$

Using the dimensional regularization, no other counterterm is generated. A φ^4 counterterm, for example, is in principle allowed by the dimensional analysis. However, it would have to be multiplied by a dimensionful parameter, which is absent in the massless case, or by the cutoff Λ, in which case it would be a powerlike divergence. We know that the dimensional regularization takes care of the potential powerlike divergences automatically, because it is a sort of dimensionless cutoff.

Exercise 8 *Compute the first nontrivial contribution to the self-energy counterterm of the massless φ_3^6 theory.*

Solution. The first correction to the self-energy is of order λ^2 and has four loops. It can be computed with the method used in exercise 5 to go from (2.63) to (2.64). The difference is that now we have to iterate the integration four times instead of two. The result is

$$\frac{\lambda^2 \mu^{4\varepsilon}}{5!} \frac{\Gamma(5-2D)\left[\Gamma\left(\frac{D}{2}-1\right)\right]^5}{(4\pi)^{2D}\Gamma\left(\frac{5D}{2}-5\right)}(k^2)^{2D-5}. \tag{2.68}$$

Extracting the divergent part, we obtain

$$Z_\varphi = 1 - \frac{4\lambda^2}{45\varepsilon(16\pi)^4} + \mathcal{O}(\lambda^3). \tag{2.69}$$

2.3 Renormalization to all orders

In renormalizable theories, which we classify in the next sections, formulas like (2.46) generalize to all orders. First, we describe what happens in detail, and later prove the claims we make. Let φ, λ and m collectively denote the fields, the couplings and the masses, respectively. Start from the classical action, and interpret it as the bare action $S_B(\varphi_B, \lambda_B, m_B)$ of the theory, which depends on the bare fields and the bare parameters, denoted by the subscript B. Then, there exist renormalization constants Z_φ, Z_{m^2} and Z_λ, which depend on λ and ε, and renormalized quantities φ, λ and m, defined by the map

$$\varphi_B = Z_\varphi^{1/2}\varphi, \qquad m_B^2 = m^2 Z_{m^2}, \qquad \lambda_B = \lambda\mu^{\sigma\varepsilon}Z_\lambda, \tag{2.70}$$

such that all the renormalized generating functionals and the renormalized correlation functions are convergent in the limit $\varepsilon \to 0$.

The renormalized generating functionals coincide with the bare generating functionals once they are written in terms of the renormalized fields and the renormalized parameters. The renormalized correlation functions are equal to the bare correlation functions, written in terms of the renormalized fields and the renormalized parameters, apart from a multiplying factor, which is specified below. In formula (2.70) σ denotes the difference between the continued and the physical dimensions of λ in units of mass, the physical dimension being the one at $\varepsilon = 0$.

Precisely, define the renormalized action S_R so that

$$S_B(\varphi_B, \lambda_B, m_B) = S_R(\varphi, \lambda, m, \mu). \tag{2.71}$$

Then, define the bare and the renormalized generating functionals Z and W by means of the formulas

$$Z_B(J_B, \lambda_B, m_B) = \int [d\varphi_B] \, e^{-S_B(\varphi_B, \lambda_B, m_B) + \int \varphi_B J_B} = e^{W_B(J_B, \lambda_B, m_B)}$$

$$= \int [d\varphi] \, e^{-S_R(\varphi, \lambda, m, \mu) + \int \varphi J} = Z_R(J, \lambda, m, \mu) = e^{W_R(J, \lambda, m, \mu)}, \quad (2.72)$$

with

$$J_B = Z_\varphi^{-1/2} J.$$

Define also bare and renormalized correlation functions, possibly connected and/or irreducible, as

$$G_B(x_1, ..., x_n, \lambda_B, m_B) = \langle \varphi_B(x_1) \cdots \varphi_B(x_n) \rangle$$
$$= Z_\varphi^{n/2} \langle \varphi(x_1) \cdots \varphi(x_n) \rangle = Z_\varphi^{n/2} G_R(x_1, ..., x_n, \lambda, m, \mu).$$

Next, using

$$\Phi_B(J_B)_x = \frac{\delta W_B(J_B)}{\delta J_B(x)} = \langle \varphi_B(x) \rangle_J$$

$$= Z_\varphi^{1/2} \langle \varphi(x) \rangle_J = Z_\varphi^{1/2} \frac{\delta W_R(J)}{\delta J(x)} = Z_\varphi^{1/2} \Phi(x),$$

perform the Legendre transforms, and construct the bare and the renormalized generating functionals Γ. The result is

$$\Gamma_B(\Phi_B, \lambda_B, m_B) = -W_B(J_B(\Phi_B)) + \int J_B(\Phi_B)\Phi_B$$

$$= -W_R(J(\Phi)) + \int J(\Phi)\Phi = \Gamma_R(\Phi, \lambda, m, \mu).$$

By definition, the map (2.70) is such that

$$\Gamma_R(\Phi, \lambda, m, \mu) < \infty,$$

in the limit $\varepsilon \to 0$, that is to say, all the irreducible diagrams must be convergent, once expressed in terms of the renormalized quantities. This fact also implies

$$Z_R(J, \lambda, m, \mu) < \infty, \quad W_R(J, \lambda, m, \mu) < \infty, \quad G_R(x_1, ..., x_n, \lambda, m, \mu) < \infty.$$

Observe that the renormalized action $S_R(\varphi, \lambda, m, \mu)$, instead, is not convergent for $\varepsilon \to 0$. Check for example, the one-loop renormalized action (2.43). We do not need to worry about this, since the classical action does not have a direct physical meaning. It is just a tool that allows us to implement what remains of the correspondence principle in quantum field theory. Renormalization, that is to say, the operation that moves the divergences away from the physical quantities (and amounts, as promised, to a change of field variables, combined with a reparametrization), does not care whether the nonphysical quantities, such as the classical action, remain meaningful or become meaningless.

Note that the renormalized sides of (2.70), (2.71), etc., depend on one quantity more than the bare sides, which is the *dynamical scale* μ. The nontrivial μ dependence of the renormalized correlation functions is the core of the renormalization-group flow, which we study later.

To prove the renormalizability of a theory to all orders, we need to prove two properties to all orders, namely, that the counterterms are local, and that they have the form of the terms already contained in the bare action S_B. We begin with the locality of the counterterms.

2.4 Locality of counterterms

Now we prove that the counterterms are local, that is to say, polynomials of the external momenta.

It is enough to discuss the irreducible diagrams, which are those that determine the Lagrangian counterterms. Once the irreducible diagrams are cured to some order, the connected and disconnected diagrams are cured to the same order.

Theorem 2 *Once the subdivergences are subtracted, the overall divergences of an irreducible diagram are local and polynomial in the masses.*

Proof. Let V_0 denote the vertices of the starting Lagrangian, and V_L the L-loop counterterms, $L \geqslant 1$. Let G_L denote an irreducible L-loop diagram built with vertices V_0, and C_L an irreducible L-loop diagram built with at least one counterterm V_N, $0 < N < L$. Note that a C_L cannot be a V_L. The counterterms V_L subtract the overall divergences of the diagrams G_L. Instead, the diagrams C_L subtract the subdivergences of G_L.

Proceeding inductively, assume that the theorem is true up to the n-th loop order included, which means that all the V_m's with $m \leqslant n$ are local and polynomial in the masses. Then, consider an irreducible diagram G_{n+1}. It corresponds to an integral over $n + 1$ loop momenta p_i, $i = 1, \ldots, n+1$. The momenta of the internal legs are linear combinations Δp_i of the p_i with coefficients ± 1. The potentially diverging contributions to the integral can only come from the integration regions where the momenta p_i are sent to infinity. The overall divergence corresponds to sending all the p_i's to infinity with the same velocity. The subdivergences correspond to sending them to infinity (still with the same velocity) under the constraint that some Δp_i's are kept fixed. Once we cure the behaviors in such integration regions, the integral is convergent, because any other integration region, such as those due to sending some Δp_i's to infinity with different velocities, are automatically covered.

From the diagrammatic point of view, keeping the momenta of some internal lines fixed, while the loop momenta are sent to infinity, amounts to cutting those internal lines, and single out a proper subdiagram. Such a subdiagram is not necessarily connected, nor irreducible. We do not need to worry about that, since the inductive assumption ensures that all the diagrams of orders $\leqslant n$ are appropriately subtracted, because, once the irreducible ones are subtracted to some order, the other diagrams are subtracted to the same order.

Observe that the subdiagrams themselves have overall divergences and subdivergences. Nevertheless, again, the inductive assumption ensures that the necessary counterterms are right there. This is actually a nontrivial fact, since we must convince ourselves that the diagrams C_L, which are built with at least one counterterm, appear in the right places, and with the right coefficients. We show that this property follows from Wick's theorem.

The examples studied before suggest that there is a direct match between the coefficients of the diagrams G_L and those of the diagrams C_L. Yet, formulas (2.56) and (2.62) tell us that we need to multiply the subtractions C_L by appropriate factors in order to match the G_L's precisely. This is the nontrivial part of the game: distribute a C_L among various G_L's, and check that the total is still equal to C_L. For example, the subtraction (e) had to be split as follows: one third for (a) and two thirds for (b). In the end, everything worked perfectly, in the cases we considered, but what is not obvious is how to promote the manipulations of those examples into a general proof to all orders. Fortunately, we are making the problem harder than it actually is. It is sufficient to change the viewpoint, to realize that all the coefficients have to match, in the end, because the subtractions ultimately belong to the integrals, not the diagrams.

To see this, we rearrange the perturbative expansion not as a sum over the diagrams G, but as a sum over the Wick contractions that lead to the diagrams (summed over the permutations of identical vertices, for reasons that we explain in a moment). We anticipated that this trick was going to be useful for some theoretical proof (although it is definitely not convenient at the practical level). Now we see how.

We know that the diagrams have complicated combinatorial factors, given by formula (1.51). Each Wick contraction has a simpler combinatorial factor, since the numerator s of (1.51) is always equal to one for bosons, and ± 1 when fermions are present. Let $\bar{G}$ denote the sum of the Wick contractions that contribute to G, and differ just by the permutations of the identical vertices. The virtue of $\bar{G}$ is that, by formula (1.51), it has a *universal* combinatorial factor, which is

$$c_{\bar{G}} = \pm \prod_i c_i^{-n_i}. \qquad (2.73)$$

We recall that n_i is the number of vertices of type i contained in G, and c_i^{-1} is the combinatorial factor that multiplies the vertex of type i in the Lagrangian. Why is (2.73) universal? Because $c_{\bar{G}}$ is equal to the product of $c_{\bar{G}_{\text{sub}}}$ for every decomposition of G into subdiagrams $\bar{G}_{\text{sub}}$. This is very convenient for the discussion of the counterterms, since it allows us to ignore the combinatorial factors altogether, and focus on the plain integrals, and their subtractions. It is not difficult to prove that the rest of the combinatorial factor c_G also factorizes among the subdiagrams, in the sense that the number of different $\bar{G}$'s contributing to G can be counted by viewing the diagram as a whole, or as a set of subdiagrams. Below we illustrate this fact in a couple of exercises.

We adopt a similar arrangement for the counterterms V_L, $L > 0$. Rather than collecting the identical contributions altogether into V_L, it is convenient to "mark" each contribution to keep track of the $\bar{G}_L$ it comes from. To illustrate this operation, we can refer to (2.34) and its splitting (2.52). Here, however, we want to do more. In (2.52) we marked the contributions to

keep track of the diagram they came from. We did not distinguish each $\bar{G}$
separately, which is why we had nontrivial coefficients floating around. Now
we want to mark each contribution to the counterterms so as to remember
the precise collection $\bar{G}$ of Wick contractions it comes from. We denote such
a contribution to V_L by $\bar{V}_L$. Recall that, after the decomposition of V_L into
a sum of $\bar{V}_L$, the combinatorial factors continue to obey the usual rules, as
shown in (2.53).

We organize the diagrams C_{n+1}, which subtract the subdivergences, ac-
cordingly. That is to say, we split each V_L as a sum of $\bar{V}_L$, and collect the
Wick contractions that just differ by the permutations of the identical vertices. Let $\bar{C}_{n+1}$ denote the contributions into which C_{n+1} is split. Clearly,
the marked counterterms $\bar{C}_{n+1}$ have the same combinatorial factor $c_{\bar{G}_{n+1}}$
of $\bar{G}_{n+1}$. Note that the same C_{n+1} is in general associated with different
diagrams G_{n+1}, because the vertices V_L used to build it, $L > 0$, are counterterms for the correlation functions (which are collections of diagrams with
the same configurations of external legs), not the single diagrams.

The "elementary" version of the theorem is the one that holds for every
$\bar{G}_{n+1}$ separately, and states that the sum

$$\bar{G}^R_{n+1} = \bar{G}_{n+1} + \sum_{\bar{C}_{n+1}(\bar{G}_{n+1})} \bar{C}_{n+1} \tag{2.74}$$

is free of subdivergences, where $\bar{C}_{n+1}(\bar{G}_{n+1})$ denotes the set of counterterms
$\bar{C}_{n+1}$ pertaining to $\bar{G}_{n+1}$. This version is the one that holds straightforwardly
(because the $\bar{C}$'s cure the divergences of the $\bar{G}$'s and the other $\bar{C}$'s in every
integration region), at the level of plain integrals, all of which are multiplied
by the same universal combinatorial factor.

Once we know that $\bar{G}^R_{n+1}$ is free of subdivergences, we also know that a
sufficient number of derivatives with respect to the masses and the external
momenta kills its overall divergence, and gives a fully convergent integral.
Therefore, the divergent part of $\bar{G}^R_{n+1}$ is local and polynomial in the masses.

Summing (2.74) on the $\bar{G}_{n+1}$'s of the same diagram G_{n+1}, we obtain the
single-diagram version of the theorem. Precisely, there exist constants $a^{C,G}_{n+1}$
such that

$$G^R_{n+1} = G_{n+1} + \sum_{C_{n+1}} a^{C,G}_{n+1} C_{n+1} \tag{2.75}$$

is free of subdivergences, where $\mathcal{C}_{n+1}$ is the set of C_{n+1}. The coefficients
$a^{C,G}_{n+1}$ can be nontrivial, because, as recalled above, the same C_{n+1} may be

associated with different G_{n+1}'s. Examples of (2.75) are (2.56) and (2.62). We infer that the divergent part of each G^R_{n+1} is local and polynomial in the masses.

Summing (2.74) on the diagrams that contribute to the same correlation function, i.e., have the same configuration of external legs, we obtain that the total

$$t = \sum_{\bar{\mathcal{G}}_{n+1}} \bar{G}_{n+1} + \sum_{\bar{\mathcal{C}}_{n+1}} \bar{C}_{n+1} = \sum_{\mathcal{G}_{n+1}} G_{n+1} + \sum_{\mathcal{C}_{n+1}} C_{n+1} = \sum_{\mathcal{G}_{n+1}} G^R_{n+1} \qquad (2.76)$$

is free of subdivergences, where the subscripts of the sums are the sets made by the addends.

Once we know that t is free of subdivergences, we also know that a sufficient number of derivatives with respect to the masses and the external momenta kills its overall divergence, and gives a fully convergent integral. Thus, the divergent part of t is polynomial in the external momenta and the masses, which ensures that the V_{n+1}'s are local and polynomial in the masses. The inductive assumption is then promoted to the order $n+1$, and so to all orders.

Comparing (2.76) to (2.75), we infer that

$$\sum_{\mathcal{G}_{n+1}} a^{C,G}_{n+1} = 1,$$

which means that the splitting of each C_{n+1} among the various diagrams G_{n+1} is such that the total remains correct. $\square$

Ultimately, the subtractions work because we manage to organize them as subtractions of integrals (which is where they belong in the first place). This allows us to forget about the combinatorics and the combinatorial factors, which always turn out to be correct.

We illustrate the $\bar{G}$-$\bar{C}$ single-integral version (2.74) of the theorem by considering a φ^3_6 two-loop diagram, together with the diagrams that subtract its subdivergences,

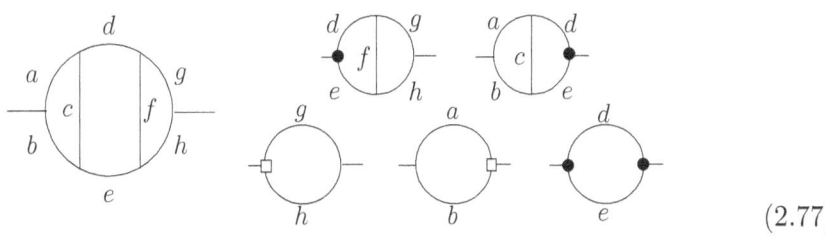

$$(2.77)$$

As above, we mark each counterterm to remember which collection $\bar{G}$ of Wick contractions it comes from. The dot and the square denote the one-loop and two-loop counterterms $\bar{V}_1$ and $\bar{V}_2$ for the vertex,

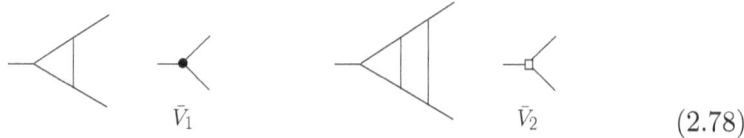

$$\bar{V}_1 \qquad\qquad\qquad \bar{V}_2 \qquad\qquad (2.78)$$

They are already equipped with the appropriate coefficients (2.73). Strictly speaking, the left dot and the right dot must be treated as different vertices.

Now we study the subdivergences, region of integration by region of integration. Let $R(u_1, ...u_k)$ denote the region where the momenta $u_1, ...u_k$ are sent to infinity, and the other ones are kept fixed. In this respect, the diagrams involving counterterms contribute to the regions where the momenta hidden by the counterterms are sent to infinity. This is because a counterterm subtracts the overall divergence of a subdiagram. So, for example, the second diagram contributes to the regions $R(a, b, c)$ and $R(a, b, c, d, e, f)$, but does not contribute to the region $R(a, b, d, e, g, h)$, because c is kept fixed there.

We have the following compensations.

1) Region $R(a, b, c)$: the first diagram is corrected by the second diagram. The other diagrams are not concerned, because in none of them the legs d, e, f, g, h are kept fixed.

2) Region $R(a, b, c, d, e, f)$: the first diagram is corrected by the second and fourth diagrams. The other diagrams are not concerned, because in none of them the legs g, h are kept fixed.

3) Region $R(a, b, c, f, g, h)$: the first diagram is corrected by the second, third and sixth diagrams. The other diagrams are not concerned, because in none of them the legs d, e are kept fixed. If you are worried that the sixth diagram is multiplied by the wrong sign (it contains two dots, so it is not a subtraction, but an addition), check carefully: two subtractions of regions $R(a, b, c, f, g, h)$ come from the second and third diagrams. This is precisely one too much, compensated by the sixth diagram.

4) The regions $R(f, g, h)$ and $R(d, e, f, g, h)$ are symmetric to the first two already considered.

5) All the other regions are trivial.

We conclude that the sum (2.77) has only overall divergences, which are local.

The argument we have illustrated in the case (2.77) generalizes to the most general diagram.

Exercise 9 *Repeat the analysis for the diagram*

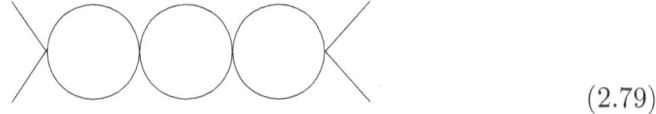

$$(2.79)$$

and its counterterms.

Exercise 10 *Count the number of $\bar{G}$'s that contribute to the diagram (a) of (2.50), and check that the result matches the one obtained by first counting the $\bar{G}$'s of the left bubble, and then the rest.*

Solution. The number of $\bar{G}$'s is $4\cdot3\cdot4\cdot3\cdot4\cdot3\cdot2$. It is obtained by counting the Wick contractions that give G, ignoring the permutations of the vertices. By definition, it is also equal to $s_G/V!$, where V is the number of vertices (three, in this case). Indeed, this is the factor that turns (2.73) into $\pm(1.51)$. Similarly, the number of $\bar{G}_{\text{sub}}$ of the left bubble is $4\cdot3\cdot4\cdot3\cdot2$. The number of ways to attach $\bar{G}_{\text{sub}}$ to the other vertex and get a $\bar{G}$ is, correctly, $4\cdot3$. The reason is that the external legs of $\bar{G}_{\text{sub}}$ are already in place. In particular, there is no factor two for the connection of the right legs of $\bar{G}_{\text{sub}}$ to the rest.

A check is obtained by exponentiating the counterterms for the simple bubble $\bar{G}_{\text{sub}}$. We need to include the permutations of the external legs, which we denote by $\bar{G}^{\text{I}}_{\text{sub}}$, $\bar{G}^{-}_{\text{sub}}$ and $\bar{G}^{\text{X}}_{\text{sub}}$. Exponentiating their counterterms $\bar{V}^{\text{a}}_{\text{sub}}$ (such that $\bar{G}^{\text{a}}_{\text{sub}} + \bar{V}^{\text{a}}_{\text{sub}} < \infty$ for every a), we get

$$\exp\left(-\frac{1}{4!}(\bar{V}^{\text{I}}_{\text{sub}} + \bar{V}^{-}_{\text{sub}} + \bar{V}^{\text{X}}_{\text{sub}})\right).$$

The factors (2.73) are included in $\bar{V}^{\text{a}}_{\text{sub}}$. Bringing the exponent down, we get

$$-\left\langle \varphi\varphi\frac{1}{4!}(\bar{V}^{\text{I}}_{\text{sub}} + \bar{V}^{-}_{\text{sub}} + \bar{V}^{\text{X}}_{\text{sub}})\frac{\varphi^4}{4!}\varphi\varphi \right\rangle.$$

When we count the Wick contractions that give the subtractions $\bar{C}$ to the diagram, we find $4\cdot3$ for attaching the right external legs to the $\varphi^4/4!$ vertex, times $4 \cdot 3$ for attaching the left external legs to a counterterm (the 4 being the number of legs of each $\bar{V}_{\text{sub}}^{\text{a}}$, and the 3 being due to the $\bar{V}_{\text{sub}}^{\text{a}}$ we can pick; the second external leg can be attached in a unique way, once the first one has been attached), times 2 for connecting the remaining internal legs. In total, we simplify the 1/4! in front of $\bar{C}_{\text{sub}}{}^1$, and get an extra $4 \cdot 3$, as we found above. Note, again, that there is no further factor 2.

This way, if we replace the $\bar{G}_{\text{sub}}$ of the left bubble with its counterterm, we have the same counting for the $\bar{C}$'s as for the $\bar{G}$'s. $\square$

Exercise 11 *Repeat the exercise for the diagram (c) of (2.50).*

Solution. Everything proceeds as above, but now the factor 2 for the connection of the internal legs of $\bar{G}_{\text{sub}}$ is there. Check that the result matches the one coming from the exponentiation of the counterterms. $\square$

The locality of the counterterms is a very general property. It does not depend on the theory, i.e., the types of fields, the forms of the propagators and the structures of the vertices, as long as a sufficient number of derivatives with respect to the external momenta, or the masses, kills the overall divergences. Any local, Lorentz invariant theory satisfies the property, and produces local counterterms. It is not required, for example, that the theory be renormalizable: nonrenormalizable theories are also cured by local counterterms, to all orders. If the vertices contain derivatives, then the integrand of (2.22) is multiplied by a certain polynomial of the momenta. Yet, it remains true that: (i) every derivative with respect to an external momentum, or a mass, lowers the potential overall degree of divergence; (ii) a sufficient number of such derivatives kills the overall degree of divergence; (iii) once the subdivergences of a diagram are properly subtracted, the overall divergences are local.

Even more, the locality of the counterterms is so general that it extends to several types of theories not considered here, including Lorentz violating and nonlocal ones.

We mentioned before that a few tricks can simplify the calculation of the divergent part of a diagram. Now we can upgrade one of those tricks. In

---

[1]Instead, the 1/4! of the vertex $\varphi^4/4!$ stays there, since it is part of the universal combinatorial factor (2.73).

general, an L-loop diagram G corresponds to an integral of the form

$$\mathcal{I}_G(k, m) = \int \prod_{i=1}^{L} \frac{\mathrm{d}^D p_i}{(2\pi)^D} \frac{P(p, k, m)}{Q(p, k, m)}, \tag{2.80}$$

instead of (2.22), where P and Q are polynomials of p, k and m. As said, nontrivial numerators P appear when the vertices contain derivatives. We know that the axioms satisfied by the analytic integral do not allow us to expand the integrand in powers of both k and m, and integrate term by term. It would be very convenient to do so, because it would allow us to efficiently isolate the overall divergent parts from the rest. We can make these operations legitimate by introducing artificial masses $\delta > 0$ in the denominators of the propagators. Specifically, let $\mathcal{I}_G^R(k, m)$ denote the subtracted integral, that is to say, the integral associated with the sum G_L^R of formula (2.75). Write

$$\mathcal{I}_G^R(k, m) = \lim_{\delta \to 0} \mathcal{I}_G^R(k, m + \delta).$$

Since $\mathcal{I}_G^R$ is equipped with the counterterms that subtract its own subdivergences, $\mathcal{I}_G(k, m + \delta)$ only has overall divergences, and those depend polynomially on k, m and δ. Separate $\mathcal{I}_G^R(k, m+\delta)$ into the sum of its divergent part $\mathcal{I}_{G\mathrm{div}}^R(k, m+\delta)$ and its convergent part $\mathcal{I}_{G\mathrm{conv}}^R(k, m+\delta)$. Since $\mathcal{I}_{G\mathrm{div}}^R(k, m+\delta)$ is a polynomial in δ, it admits a smooth limit for $\delta \to 0$, which we denote by $\bar{\mathcal{I}}_{G\mathrm{div}}^R(k, m)$. Then, $\mathcal{I}_{G\mathrm{conv}}^R$ also admits a smooth limit for $\delta \to 0$, because the sum $\mathcal{I}_{G\mathrm{div}}^R(k, m + \delta) + \mathcal{I}_{G\mathrm{conv}}^R(k, m + \delta)$ must tend to $\mathcal{I}_G^R(k, m)$. Thus, we can write

$$\mathcal{I}_G^R(k, m) = \bar{\mathcal{I}}_{G\mathrm{div}}^R(k, m) + \lim_{\delta \to 0} \mathcal{I}_{G\mathrm{conv}}^R(k, m + \delta).$$

The second term on the right-hand side has no poles for $\varepsilon \to 0$, so it is convergent even after taking the limit $\delta \to 0$. Finally, the divergent part $\mathcal{I}_{G\mathrm{div}}^R(k, m)$ of the subtracted diagram G_L^R can be calculated as

$$\mathcal{I}_{G\mathrm{div}}^R(k, m) = \bar{\mathcal{I}}_{G\mathrm{div}}^R(k, m) = \lim_{\delta \to 0} \mathcal{I}_{G\mathrm{div}}^R(k, m + \delta). \tag{2.81}$$

We stress again that it is not legitimate to expand the integrand of $\mathcal{I}_G(k, m)$ in powers of both k and m, and then integrate term by term. However, these operations are legitimate on $\mathcal{I}_G^R(k, m + \delta)$, as long as δ is nonzero. Formula (2.81) tells us that when we set δ back to zero, we recover the full divergent part of $\mathcal{I}_G^R(k, m)$.

The upgraded trick is convenient in several situations. If we expand in powers of the external momenta and the masses, and use the dimensional regularization, we obtain linear combinations of (simpler) integrals that have vanishing degrees of divergence: every powerlike-divergent integral can be dropped, since its divergent part is killed by the $\delta \to 0$ limit of (2.81). In the massless theories, the artificial mass allows us to compute the divergent parts by expanding in powers of the external momenta k.

Exercise 12 *Prove that, if we use the dimensional regularization, a local quantum field theory in odd dimensions has no nontrivial L-loop overall divergences, if L is also odd.*

Solution. The integrals have the form (2.80). We insert artificial masses δ in the denominators, then expand in powers of the masses m and the external momenta k. In the end, the overall divergences are given by expressions of the form

$$\int \prod_{i=1}^{L} \frac{\mathrm{d}^D p_i}{(2\pi)^D} \frac{(p_{i_1})_{\mu_1} \cdots (p_{i_n})_{\mu_n}}{Q'(p^2)}, \qquad (2.82)$$

where $(p_i)_\mu$ is the μ-th component of the i-th loop momentum, and the denominator is a polynomial in the squared momenta p_i^2 and Δp_j^2. If both d and L are odd, n must be odd, to have a vanishing degree of divergence: otherwise, the integral is either convergent or powerlike divergent (and then its divergent part disappears in the limit $\delta \to 0$). However, if n is odd, the integral (2.82) is odd under the transformation $p_i \to -p_i$, so its overall divergent part vanishes. $\square$

2.5   Power counting

The renormalizability of a theory can be established by means of a simple dimensional analysis, called *power counting*.

Consider a d-dimensional theory of interacting bosons φ and fermions ψ. We assume that the bosonic fields have propagators $P_B(k)$ that behave like $\sim 1/k^2$ for large momenta k. By this we mean that behaviors such as

$$P_B(k) \sim \sum_n c_n \frac{k_{\mu_1} \cdots k_{\mu_{2n}}}{(k^2)^{n+1}} \qquad (2.83)$$

are also allowed. Similarly, we assume that the fermionic fields have propagators $P_F(k)$ that behave like $\sim k_\mu/k^2$, or, more generally,

$$P_F(k) \sim \sum_n d_n \frac{k_{\mu_1} \cdots k_{\mu_{2n+1}}}{(k^2)^{n+1}}, \tag{2.84}$$

for large momenta. Such behaviors tell us that the dimensions of the bosonic and the fermionic fields are $(d-2)/2$ and $(d-1)/2$, respectively.

More generally, consider fields χ_a of dimensions $d/2 - a$ with propagators that behave like

$$P_a(k) \sim \sum_n e_n \frac{k_{\mu_1} \cdots k_{\mu_n}}{(k^2)^{n/2+a}}$$

at large momenta, where a is integer or half-integer, and n is even or odd, respectively. We are not making assumptions about the sign of a, nor the statistics of χ_a.

Let m collectively denote the masses of the fields. Let n_{iB}, n_{iF}, n_{ia} denote the numbers of legs of types B, F and a of the ith vertex. Assume that each vertex is a monomial $V_i(k)$ in the momenta (multiplied by the appropriate couplings). We can always split a polynomial vertex as a linear combination of the $V_i(k)$'s and treat the monomials separately. Let δ_i denote the dimension of $V_i(k)$ in units of mass.

Consider a diagram G with E_B, E_F, E_a external legs and I_B, I_F, I_a internal legs of the various types, v_i vertices of the i-th type and L loops. We have, from (2.18)

$$L - I_B - I_F - \sum_a I_a + V = 1, \qquad V = \sum_i v_i. \tag{2.85}$$

We denote the external momenta with k and the loop momenta with p. The integral associated with G has the form

$$\mathcal{I}_G(k, m) = \int \frac{\mathrm{d}^{Ld} p}{(2\pi)^{Ld}} \prod_{j=1}^{I_B} P_{Bj}(p, k, m) \prod_{l=1}^{I_F} P_{Fl}(p, k, m) \times$$

$$\times \prod_a \prod_{j_a=1}^{I_a} P_{aj_a}(p, k, m) \prod_i \prod_{l_i=1}^{v_i} V_{il_i}(p, k, m), \tag{2.86}$$

where the indices j, l, j_a and l_i of P_{Bj}, P_{Fl}, P_{aj_a} and V_{il_i} label the propagators and the vertices.

Now, rescale k and m to λk and λm. It is convenient to rescale the loop momenta as well, which is just a change of variables in the integral. Then $\mathcal{I}_G(k, m)$ rescales by a power of λ that is equal to the dimension of $\mathcal{I}_G$ in units of mass, which is

$$[\mathcal{I}_G] = Ld - 2I_B - I_F - 2\sum_a aI_a + \sum_i v_i \delta_i. \tag{2.87}$$

Since the overall divergences are local, once the subdivergences have been subtracted away, we infer that they are a polynomial of degree $w_G \leqslant [\mathcal{I}_G]$ in the external momenta and the masses.

Now, count the bosonic legs attached to the vertices: they can exit the diagram or be connected to other bosonic internal legs, so

$$E_B + 2I_B = \sum_i v_i n_{iB}.$$

Similarly, counting the fermionic legs and the χ_a legs, we obtain

$$E_F + 2I_F = \sum_i v_i n_{iF}, \qquad E_a + 2I_a = \sum_i v_i n_{ia}.$$

Using (2.85)-(2.87) and $w_G \leqslant [\mathcal{I}_G]$ we get

$$w_G \leqslant d(E_B, E_F, E_a) + \sum_i v_i \left[\delta_i - d(n_{iB}, n_{iF}, n_{ia})\right],$$

where

$$d(x, y, z_a) \equiv d - \frac{d-2}{2}x - \frac{d-1}{2}y - \sum_a \frac{d-2a}{2} z_a.$$

We see that if all the vertices satisfy

$$\delta_i \leqslant d(n_{iB}, n_{iF}, n_{ia}) \tag{2.88}$$

then all the counterterms satisfy the same inequality, namely,

$$w_G \leqslant d(E_B, E_F, E_a). \tag{2.89}$$

In other words, if the classical Lagrangian includes all the vertices that satisfy (2.88), then the divergent parts of all the diagrams can be subtracted away by renormalizing the couplings, the fields and the masses. The conditions (2.88) define a theory that is *renormalizable by power counting*.

Instead, if the Lagrangian contains some vertex $\bar{v}$ that does not satisfy (2.88), the diagrams that contain $\bar{v}$ can have arbitrarily large degrees of divergence. In general, in that case, it is necessary to add infinitely many new vertices and couplings to the Lagrangian, if we want to subtract the divergences by means of redefinitions of the fields and the parameters. This kind of theory is called *nonrenormalizable*.

The renormalizable theories where all the δ_i's are strictly smaller than $d(n_{iB}, n_{iF}, n_{ia})$ are called *super-renormalizable*.

It is easy to check that the requirement (2.88) is equivalent to demanding that all the Lagrangian terms have coefficients of nonnegative dimensions in units of mass. Indeed, the dimension of the coupling λ_i that multiplies the i-th vertex is

$$[\lambda_i] = d - \frac{d-2}{2} n_{iB} - \frac{d-1}{2} n_{iF} - \sum_a \frac{d-2a}{2} n_{ia} - \delta_i = d(n_{iB}, n_{iF}, n_{ia}) - \delta_i \geqslant 0.$$

Thus, a theory is renormalizable by power counting if it contains no parameters of negative dimensions (and the propagators are well behaved). This conclusion can be derived more quickly as follows. At the level of the Lagrangian, a counterterm, being local, must have the structure

$$\left(\prod \lambda \right) \partial^r \varphi^{n_B} \psi^{n_F} \prod_a \chi_a^{n_{ia}}. \tag{2.90}$$

The coefficient is a certain product of couplings and masses. We do not need to specify where the derivatives act in (2.90), since it is not important for our discussion. Now, the dimension of (2.90) must be equal to d. If the theory contains no parameters of negative dimensions, we must have

$$r + n_{iB} \frac{d-2}{2} + n_{iF} \frac{d-1}{2} + \sum_a \frac{d-2a}{2} n_{ia} \leqslant d,$$

which is equivalent to (2.88). On the other hand, if the theory contains parameters $\tilde{\lambda}$ of negative dimensions, then arbitrarily large powers $h_{\tilde{\lambda}}$ of $\tilde{\lambda}$ can multiply the counterterm, and we just have an inequality of the form

$$r + n_{iB} \frac{d-2}{2} + n_{iF} \frac{d-1}{2} + \sum_a \frac{d-2a}{2} n_{ia} \leqslant d - \sum_{\tilde{\lambda}} [\tilde{\lambda}] h_{\tilde{\lambda}}, \tag{2.91}$$

which can violate (2.88), depending on the other parameters that multiply the counterterm.

Now, it should be kept in mind that, in general, in renormalization theory, the following "no-miracle" principle applies:

all the counterterms that are not a priori forbidden are generated by renormalization.

A counterterm can be forbidden on general grounds by power counting, gauge symmetries, and external symmetries. If it is not forbidden, there is practically no hope that it will not be generated as the divergent part of a diagram with an appropriate set of external legs. In other words, no miraculous cancellations should be expected. If (2.88) can be violated, it will be. For example, it is violated by the counterterms (2.90) where no parameters of positive dimensions compensate for the $\tilde{\lambda}$ couplings, in which case (2.91) holds with the equality sign. This implies that infinitely many new types of counterterms are effectively generated, and the theory is nonrenormalizable.

It is important to stress that the propagators must have the right behaviors for large momenta. For example, the Proca vectors of formula (1.89) are in general not renormalizable, when interactions are present. Indeed, the propagator (1.91) contains a term $\sim p_\mu p_\nu/(m^2 p^2)$ that prevails over $\delta_{\mu\nu}/p^2$ at large momenta. This forces us to treat the field as a χ_a-field with $a = 0$, which means that its dimension, from the viewpoint of power counting, is equal to $d/2$. The fields of such a dimension can appear at most quadratically in a local field theory, so they cannot have renormalizable self-interactions.

The gauge fields can instead be included consistently, because their propagators are well behaved, once they are properly introduced. This is not an obvious task, which is why the gauge theories deserve a special treatment. We devote the final chapters of this book to prove their renormalizability.

Particular (scalar) fields of dimension $d/2$ can be useful as auxiliary fields. For example, in the massless φ^4_4 theory we can introduce an auxiliary field σ of dimension 2 and replace the φ^4-vertex with

$$\mathcal{L}'_I = \frac{1}{2}\sigma^2 + i\mu^{\varepsilon/2}\alpha\sigma\varphi^2, \qquad (2.92)$$

where $\alpha = \sqrt{\lambda/12}$. The integral over σ can be performed exactly, by means of a translation $\sigma' = \sigma + i\mu^{\varepsilon/2}\alpha\varphi^2$, which brings $\mathcal{L}'_I$ to the form

$$\mathcal{L}'_I = \frac{1}{2}\sigma'^2 + \frac{\lambda\mu^\varepsilon}{4!}\varphi^4.$$

At this point, the field σ' decouples and can be dropped, so the modified theory is equivalent to the φ^4_4 theory. Sometimes it can be useful to work out the Feynman rules and the diagrams from (2.92). In that case, σ has a propagator equal to 1, so it is a χ_a field with $a = 0$. The renormalizability by power counting still works. We just need to add an extra vertex φ^4 to the Lagrangian, because it is allowed by power counting. We multiply it by an independent coupling λ' and treat α as an independent coupling as well. In total, the renormalized Lagrangian $\mathcal{L}'_I$ reads

$$\mathcal{L}'_{\text{IR}} = \frac{Z_\sigma}{2}\sigma^2 + i\mu^{\varepsilon/2}\alpha Z_\alpha Z_\sigma^{1/2} Z_\varphi \sigma \varphi^2 + \frac{\lambda'\mu^\varepsilon Z'_\lambda Z_\varphi^2}{4!}\varphi^4 = \frac{1}{2}\sigma'^2 + \frac{\lambda\mu^\varepsilon Z_\lambda}{4!}Z_\varphi^2\varphi^4$$

where now $\sigma' = Z_\sigma^{1/2}\sigma + i\mu^{\varepsilon/2}\alpha Z_\alpha Z_\varphi\varphi^2$ and $\lambda Z_\lambda = \lambda' Z'_\lambda + 12\alpha^2 Z_\alpha$. The theory is equivalent to the ordinary massless φ_4^4 theory with the coupling $\lambda = \lambda' + 12\alpha^2$.

The no-miracle principle also implies that a renormalizable theory must contain *all* the Lagrangian terms that are not a priori forbidden. Indeed, assume that for some reason we start with a Lagrangian with some missing vertex $\bar{v}$. A divergent contribution $\bar{c}$ of the same form will be generated by renormalization. To subtract it, it is necessary to go back to the classical Lagrangian and add $\bar{v}$, multiplied by a new coupling $\bar{\lambda}$. Once that is done, it is possible to remove $\bar{c}$ by making a redefinition of $\bar{\lambda}$. We see that, because of renormalization, we are not free to choose the theory we like. Most classical theories make no sense at the quantum level, either because they do not contain enough vertices, the renormalizable ones, or because they contain nonrenormalizable vertices. Renormalization either guides us towards the right theory, or blows up into (2.40). In this sense, it provides a way to *select* the theories.

Sometimes, the parameters of zero dimension are called *marginal*, those of positive dimensions *relevant* and those of negative dimensions *irrelevant*. This terminology refers to the low-energy behavior of the theory. For example, the parameters of negative dimensions multiply Lagrangian terms of dimensions larger than d, which are indeed negligible in the low-energy limit. Instead, the parameters of positive dimensions multiply the terms that are more "relevant" at low energies. This terminology will not be used further in this book.

2.6   Renormalizable theories

The list of renormalizable theories depends on the spacetime dimension d. We start from four dimensions, where

$$d(n_B, n_F) = 4 - n_B - \frac{3}{2} n_F.$$

By locality, $d(n_B, n_F)$ must be nonnegative, so n_B can be at most 4 and n_F can be at most 2. Note that n_F must be even. We have the following possibilities

(n_B, n_F)	$(1,0)$	$(2,0)$	$(3,0)$
$d(n_B, n_F)$	3	2	1
Lagrangian terms	φ	$\varphi^2, \varphi\partial\varphi, (\partial\varphi)^2$	$\varphi^3, \varphi^2\partial\varphi$
(n_B, n_F)	$(4,0)$	$(0,2)$	$(1,2)$
$d(n_B, n_F)$	0	1	0
Lagrangian terms	φ^4	$\bar{\psi}\psi, \bar{\psi}\partial\psi$	$\varphi\bar{\psi}\psi$

The notation is symbolic, in the sense that we do not pay attention to where the derivatives act and how the indices are contracted. The most complicated bosonic interaction is φ^4, and the most complicated scalar-fermion interaction is the Yukawa vertex $\varphi\bar{\psi}\psi$. No fermion self-interaction is allowed.

The structure of the most general four-dimensional Lorentz invariant Lagrangian of scalar fields φ, vectors A and fermions ψ is

$$\mathcal{L}_4 = \frac{1}{2}(\partial_\mu A_\nu)^2 - \frac{\xi}{2}(\partial_\mu A_\mu)^2 + \frac{m_A^2}{2} A_\mu^2 + \frac{1}{2}(\partial_\mu\varphi)^2 + \frac{m_s^2}{2}\varphi^2$$

$$+ \bar{\psi}\partial\!\!\!/\psi + m_f\bar{\psi}\psi + \lambda_{1s}\varphi + \frac{\lambda_{3s}}{3!}\varphi^3 + \frac{\lambda_{4s}}{4!}\varphi^4 + \frac{g_3}{3!}A_\mu^2(\partial_\nu A_\nu)$$

$$+ \frac{g_3'}{3!}A_\mu A_\nu \partial_\nu A_\mu + \frac{g_4}{4!}(A_\mu^2)^2 + Y_s\varphi\bar{\psi}\psi + Y_A A_\mu \bar{\psi}\gamma_\mu\psi + \frac{g_{2s}}{2}\varphi(\partial_\mu A_\mu)$$

$$+ \frac{g_{3s}}{3!}\varphi A_\mu^2 + \frac{g_{3s}'}{3!}\varphi^2\partial_\nu A_\nu + \frac{g_{3v}''}{3!}A_\mu\varphi\partial_\mu\varphi + \frac{\lambda_{4sv}}{4!}\varphi^2 A_\mu^2,$$

up to total derivatives, where $\xi \neq 1$, plus fermionic terms equal to the listed ones with $\psi \to \gamma_5\psi$, where γ_5 is the product of the four γ matrices γ_μ. Some terms appear to differ by total derivatives, but they are actually independent when vectors of more types are present.

Let us assume that we can diagonalize the vectors from the scalars (i.e., $g_{2s} = 0$). At $\xi \neq 1$ the vector propagator behaves correctly for large momenta, from the point of view of renormalization, even when the mass m_A

does not vanish. However, we anticipate that a theory with $\xi \neq 1$ is not unitary, that is to say, it propagates unphysical degrees of freedom. On the other hand, at $\xi = 1$, $m_A \neq 0$ the vector A_μ is of the Proca type, which has a bad behavior for large momenta. Finally, at $\xi = 1$, $m_A = 0$, the vector propagator does not exist. This is the case of gauge theories, which will be treated in the next chapters. We will show that, after a suitable "gauge-fixing", the propagator becomes well behaved, and the theory can be proved to be both unitary and renormalizable by power counting.

Simple examples of renormalizable theories in four dimensions are the φ_4^4 theory (2.47), which is renormalized by (2.46) in the form (2.44), and the massless Yukawa theory (2.17). Its bare action

$$S_B(\varphi_B, \psi_B) \equiv \int d^D x \left(\frac{1}{2}(\partial_\mu \varphi_B)^2 + \lambda_B \frac{\varphi_B^4}{4!} + \bar{\psi}_B \slashed{\partial} \psi_B + g_B \varphi_B \bar{\psi}_B \psi_B \right) \quad (2.93)$$

is renormalized by the map

$$\varphi_B = Z_\varphi^{1/2} \varphi, \qquad \lambda_B = \lambda \mu^\varepsilon Z_\lambda, \qquad \psi_B = Z_\psi^{1/2} \psi, \qquad g_B = g \mu^{\varepsilon/2} Z_g,$$

which gives the renormalized action

$$S_R(\varphi, \psi) \equiv \int d^D x \left(\frac{Z_\varphi}{2}(\partial_\mu \varphi)^2 + \lambda \mu^\varepsilon Z_\lambda Z_\varphi^2 \frac{\varphi^4}{4!} \right.$$
$$\left. + Z_\psi \bar{\psi} \slashed{\partial} \psi + g \mu^{\varepsilon/2} Z_g Z_\varphi^{1/2} Z_\psi \varphi \bar{\psi} \psi \right). \quad (2.94)$$

We can use this example to illustrate what happens when we start from a theory with some missing vertex. Assume that we "forget" the φ^4 vertex and start with the Lagrangian

$$\mathcal{L} = \frac{1}{2}(\partial_\mu \varphi)^2 + \bar{\psi} \slashed{\partial} \psi + g \mu^{\varepsilon/2} \varphi \bar{\psi} \psi. \quad (2.95)$$

Then, consider the one-loop diagram

$$(2.96)$$

and its permutations, where the dashed lines denote the scalars and the continuous ones denote the fermions. It is easy to check that the divergent part of (2.96) is nonvanishing. Thus, the theory (2.95) is not renormalizable as it stands. The missing φ^4 vertex must be added to the classical Lagrangian, after which (2.95) becomes (2.17). Then, the redefinition of the φ^4 coupling can remove the divergent part of the diagram (2.96). We conclude that the theory (2.95) makes no sense at the quantum level, although it is a perfectly meaningful classical theory. Only (2.17) makes sense.

In three spacetime dimensions, we have

$$d(n_B, n_F) = 3 - \frac{1}{2}n_B - n_F - \frac{3 - 2a}{2}n_a,$$

so $n_B \leqslant 6$ and $n_F \leqslant 2$. The most complicated bosonic interaction is φ^6. Again, no fermionic self-interaction is admitted. We can have interactions among two bosons and two fermions, such as $\varphi^2 \bar{\psi}\psi$, $A_\mu^2 \bar{\psi}\psi$, etc.

We have included a field χ_a, because there exist interesting vectors with $a = 1/2$ and propagators $\sim k_\mu/k^2$ (Chern-Simons fields). Their kinetic term reads

$$\mathcal{L}_{CS} = \frac{i}{2}A_\mu \partial_\rho A_\nu \varepsilon_{\mu\nu\rho}.$$

The condition on the number of $\chi_{1/2}$ external legs is $n_{1/2} \leqslant 3$. Vertices

$$A_\mu A'_\rho A''_\nu \varepsilon_{\mu\nu\rho} \tag{2.97}$$

are allowed (with vectors of several types). Apart from constraints coming from the statistics, the Chern-Simons fields behave like fermions ψ.

Summarizing,

(n_B, n_F)	$(2,0)$	$(4,0)$	$(6,0)$	$(2,1)$
$d(n_B, n_F)$	2	1	0	1
Lagrangian terms	$(\partial\varphi)^2$	$\partial\varphi^4$	φ^6	$\partial\varphi^2\psi$
(n_B, n_F)	$(4,1)$	$(0,2)$	$(2,2)$	$(0,3)$
$d(n_B, n_F)$	0	1	0	0
Lagrangian terms	$\varphi^4\psi$	$\psi\partial\psi$	$\varphi^2\psi^2$	ψ^3

where φ can stand for scalar fields and ordinary vector fields, while ψ can stand for fermions and Chern-Simons vectors. We have listed only the Lagrangian terms that have the largest numbers of fields and derivatives. The derivatives can act anywhere. The missing terms are obtained from the listed ones by suppressing fields and/or derivatives.

In six dimensions

$$d(n_B, n_F) = 6 - 2n_B - \frac{5}{2}n_F,$$

which implies $n_B \leqslant 3$, $n_F \leqslant 2$. Moreover, $n_F = 2$ implies that n_B must vanish, so the fermions are free. It is said that they "decouple", and so can be ignored for our present purposes. Hence, the only allowed interaction is φ^3. However, the theory φ_6^3 is not physically interesting, since the potential φ^3 is not bounded from below. Vectors decouple also, by Lorentz invariance, unless parity is violated, in which case a self-interaction (2.97) is allowed. Yet, this interaction is not allowed by gauge invariance. Moreover, non-Proca, non-gauge vectors are not unitary. In five dimensions the situation is the same as in six. In dimensions greater than six all the fields are free.

We see that only dimensions smaller than or equal to four admit physically acceptable interacting renormalizable theories. Unfortunately, gravity is not renormalizable in four dimensions. It is described by a spin-2 field (a symmetric tensor) $h_{\mu\nu}$, which has derivative interactions of the form

$$\sim \kappa^n h^n \partial h \partial h,$$

which involve a coupling κ, the Newton constant, of dimension -1 in units of mass.

Newton's constant is dimensionless in two dimensions, which suggests that gravity is power counting renormalizable there. However, it can be shown that general relativity in less than four dimensions contains no propagating graviton. In less than three dimensions gauge vectors have no propagating degree of freedom either. We discover that the renormalizable interactions are very few, which means that renormalizability is an extremely powerful criterion to select the theories. It is so restrictive that four is the only number of dimensions where the renormalizable theories are diverse enough to explain the three interactions of nature besides gravity. However, the fact that gravity is not renormalizable by power counting in $d > 2$ suggests that a more profound understanding of renormalization must exist.

The renormalizable theories are the theories where the subtraction algorithm achieves its goal of removing all the divergences by redefining the fields and a finite number of independent parameters. The theories studied

in this book are renormalizable according to the mere rules of power counting. However, it should be kept in mind that there may exist theories that
are renormalizable by other criteria (global or discrete symmetries, selection
rules, large N expansions, etc.), often in combination with power counting.

Exercise 13 *Compute the one-loop renormalization of the massless scalar-
fermion theory (2.17).*

   Solution. The nontrivial divergent diagrams are

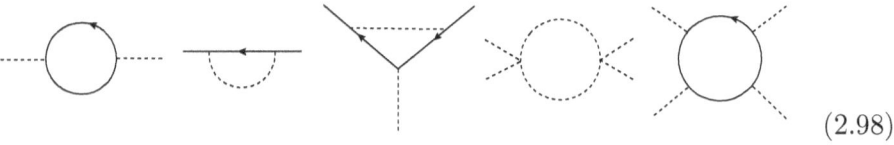

$$(2.98)$$

The calculation can be simplified by means of the tricks explained previously
in this chapter. Note that the last diagram gives 6 identical contributions:
a factor 2 comes from the orientation of the loop and a factor 3 comes from
the permutations of the external legs. We find

$$Z_\varphi = 1 - \frac{4g^2}{(4\pi)^2\varepsilon}, \qquad Z_\psi = 1 - \frac{g^2}{(4\pi)^2\varepsilon}, \qquad Z_g = 1 + \frac{5g^2}{(4\pi)^2\varepsilon},$$

$$\lambda Z_\lambda = \lambda + \frac{1}{(4\pi)^2\varepsilon}(3\lambda^2 + 8g^2\lambda - 48g^4).$$

There is no diagram of order $g^2\lambda$, so $\lambda Z_\lambda Z_\varphi^2$ does not contain such a type
of contribution. Note that, in general, when the theory contains more cou-
plings, the vertex renormalization constants, like Z_λ in this example, may
contain a negative power of the coupling. Since λZ_λ is certainly polynomial
(for every truncation of the perturbative expansion), it is better to rewrite
λZ_λ as $\lambda + \Delta_\lambda$, where Δ_λ collects the counterterms and is also polynomial.

Exercise 14 *Compute the one-loop renormalization of the four-fermion theory (1.106) in two dimensions, where ψ is a multiplet made of N copies of
the basic spinor doublet.*

   Solution. This theory is, in some respects, similar to the φ^4 theory
in four dimensions. The Feynman rules are (1.107) with $\lambda \to \lambda\mu^\varepsilon$, where
$\varepsilon = 2 - D$. There is no wave-function renormalization at one loop. The mass

renormalization is given by a tadpole diagram, which turns out to be equal to $-\lambda(2N-1)m\mu^\varepsilon/2$ times (2.6), where $2N-1$ comes from evaluating the fermion loop. Expanding the left-hand side of (2.7) around two dimensions, we get

$$Z_m = 1 - \frac{(2N-1)\lambda}{4\pi\varepsilon}.$$

The vertex renormalization is given by the diagrams

$$\frac{1}{2}\ \text{}\ -\ \text{}\ +\ \text{}$$

where the combinatorial factors and the signs due to the fermion exchanges and the fermion loops are written explicitly. Observe that the first diagram is not multiplied by (-1), since it does not contain a closed fermion loop. Instead, the third diagram has a plus sign, since the factor (-1) due to the closed fermion loop is compensated by another factor (-1) due to the permutation of two external identical fermions. Using the two-dimensional identity

$$(\gamma_\mu)^{\alpha\beta}(\gamma_\mu)^{\gamma\delta} - (\gamma_\mu)^{\gamma\beta}(\gamma_\mu)^{\alpha\delta} = -2(\delta^{\alpha\beta}\delta^{\gamma\delta} - \delta^{\alpha\delta}\delta^{\gamma\beta}),$$

where γ_μ are the first two Pauli matrices, we obtain

$$Z_\lambda = 1 - \frac{(N-1)\lambda}{2\pi\varepsilon}. \tag{2.99}$$

Exercise 15 *Write the Lagrangian of the previous exercise in the equivalent form*

$$\mathcal{L} = \bar{\psi}(\partial\!\!\!/ + m)\psi + \sqrt{\frac{\lambda}{2}}\mu^{\varepsilon/2}\sigma\bar{\psi}\psi + \frac{1}{2}\sigma^2,$$

having introduced an auxiliary field σ. Renormalize the theory in this form at one loop, and check the results already found.

Solution. The divergent one-loop diagrams are the first three of the list (2.98), plus a tadpole (the fermion loop with one external leg σ), plus the diagrams made by two fermion lines connected by two σ lines. The last ones could in principle turn on a counterterm proportional to $(\bar{\psi}\psi)^2$, but ultimately they do not. Indeed, they can be arranged into pairs (the σ lines can cross or not), and it is easy to see that the divergent parts mutually cancel within each pair. The rest of the calculation is straightforward (and much simpler than before). The result is the renormalized Lagrangian

$$\mathcal{L}_R = \bar{\psi}\partial\!\!\!/\psi + \left(1 + \frac{\lambda}{4\pi\varepsilon}\right)\left(m + \sqrt{\frac{\lambda}{2}}\mu^{\varepsilon/2}\sigma\right)\bar{\psi}\psi$$

$$+ \frac{1}{2}\sigma^2\left(1 + \frac{\lambda N}{2\pi\varepsilon}\right) + \sqrt{\frac{\lambda}{2}}\frac{mN}{\pi\varepsilon}\mu^{-\varepsilon/2}\sigma. \qquad (2.100)$$

Integrating σ away, we retrieve the results of the previous exercise. Note that to have (2.100) real, the coupling λ must be positive. This is the reason why we have put a minus sign in front of the four fermion vertex of the Lagrangian (1.106). If the fields were bosonic, that minus sign would be wrong. For the reason just explained, it is the right sign for fermionic fields. □

The last two exercises teach us that, if we make a change of field variables, the theory remains renormalizable, but the renormalization organizes itself in a different way. For example, the Lagrangian (1.106) has nontrivial renormalization constants for m and λ, at one loop, while (2.100) also has a renormalization constant for σ, and contains a σ linear term. The two renormalized Lagrangians are mapped into each other by a renormalized change of field variables.

Exercise 16 *Find the renormalized change of field variables that maps the Lagrangian (2.100) into the renormalized version of (1.106).*

Solution. It is

$$\sigma = \left(1 - \frac{\lambda N}{4\pi\varepsilon}\right)\sigma' - \left(1 - \frac{(2N-1)\lambda}{4\pi\varepsilon}\right)\sqrt{\frac{\lambda}{2}}\mu^{\varepsilon/2}\bar{\psi}\psi$$

$$- \sqrt{\frac{\lambda}{2}}\mu^{-\varepsilon/2}\left(1 - \frac{\lambda N}{2\pi\varepsilon}\right)\frac{mN}{\pi\varepsilon},$$

plus higher orders. Indeed, (2.100) is equal to

$$\bar{\psi}\partial\!\!\!/\psi + mZ_m\bar{\psi}\psi - \frac{\lambda}{4}\mu^{\varepsilon}Z_\lambda(\bar{\psi}\psi)^2 + \frac{1}{2}\sigma'^2,$$

plus a constant, plus higher orders. This is the renormalized version of (1.106) at one loop, plus a quadratic term that decouples (and is not renormalized).

2.7   Composite fields

Composite fields are products of elementary fields and their derivatives in the same spacetime point. Because of this, they are local. They are also called "operators", or "composite operators", although strictly speaking no operator is involved in the functional-integral approach. The simplest example is the composite field $\varphi^2(x)$ in a scalar theory. Sometimes it is useful to consider local functionals as well, that is to say, integrals of composite fields over the whole of spacetime. Local functionals are also called integrated composite fields, or integrated operators.

The renormalization of a composite field is in general not related in an obvious way to the renormalization of its component fields, and has to be calculated anew. For example, the renormalization of φ^2 is unrelated to the renormalization of φ, in the sense that the correlation functions that contain an insertion of φ^2 can be made convergent with the help a new renormalization constant, Z_{φ^2}, which has no relation with Z_φ. The correlation functions that contain more insertions of φ^2 can be made convergent by means of subtractions that are independent of both Z_{φ^2} and Z_φ.

Let us recall that Z_φ renormalizes the divergences of the correlation functions where the scalar fields are inserted at different spacetime points, e.g.,

$$\langle \varphi(x)\varphi(y)\varphi(z)\varphi(w)\rangle. \tag{2.101}$$

We may want to consider different types of correlation functions, possibly containing insertions of φ^2, such as

$$\langle \varphi^2(x)\varphi(y)\varphi(z)\rangle, \qquad \langle \varphi^2(x)\varphi^2(y)\rangle. \tag{2.102}$$

We know that the correlation functions have to be meant as distributions. In a distribution it often makes no sense to study the limit of coincident points. Therefore, the first of (2.102) is *not* the $w \to x$ limit of (2.101), and the second of (2.102) is not the $z \to x$, $w \to y$ limit of (2.101). They are just different distributions.

We must distinguish bare and renormalized composite fields. The bare composite fields are denoted by $\mathcal{O}_B$. They are just the products of the bare factors. For example, the bare operator $\varphi^2(x)$ is just the product of two bare scalar fields in x, i.e., $\varphi_B^2(x)$. The renormalized composite fields are denoted by $\mathcal{O}_R$, or by writing the composite field in between square brackets, such as

$[\varphi^2(x)]$, to distinguish it from $\varphi^2(x)$, which is the product of the renormalized fields.

Bare and renormalized operators are related by new renormalization constants $Z_\mathcal{O}$:

$$\mathcal{O}_\mathrm{B} = Z_\mathcal{O}\mathcal{O}_\mathrm{R}. \qquad (2.103)$$

For example, we have $\varphi_\mathrm{B}^2 = Z_{\varphi^2}[\varphi^2]$. On the other hand, we know that $\varphi_\mathrm{B} = Z_\varphi^{1/2}\varphi$, hence

$$[\varphi^2] = Z_{\varphi^2}^{-1}Z_\varphi\varphi^2. \qquad (2.104)$$

This formula emphasizes that the renormalized operator $[\varphi^2]$ does not coincide with the square of the renormalized field φ, unless $Z_{\varphi^2} = Z_\varphi$, which is in general not true.

Thus, the renormalized three-point function of (2.102) reads

$$\langle[\varphi^2(x)]\varphi(y)\varphi(z)\rangle = Z_{\varphi^2}^{-1}Z_\varphi^2\langle\varphi^2(x)\varphi(y)\varphi(z)\rangle. \qquad (2.105)$$

Here $Z_{\varphi^2}^{-1}$ cancels the extra divergences due to the coincident points of $\varphi^2(x)$.

A composite field $\mathcal{O}(\varphi)$ can be described as a vertex, and included in the Lagrangian by coupling it to an external source, which we denote by L. This way, the $\mathcal{O}(\varphi)$ insertions into a correlation function can be studied as functional derivatives with respect to L.

In general, the renormalization of composite fields is not just multiplicative, because different composite fields can "mix" under renormalization. For this reason, it is convenient to treat them all at once.

Let $\mathcal{O}_\mathrm{B}^I = \mathcal{O}^I(\varphi_\mathrm{B})$ denote a basis of bare composite fields, and $\mathcal{O}_\mathrm{R}^I$ a basis of renormalized composite fields. Let L_B^I and L^I denote the bare and the renormalized sources coupled to them.

At the bare level, we just need to add

$$-L_\mathrm{B}^I\mathcal{O}^I(\varphi_\mathrm{B}) \qquad (2.106)$$

to the bare Lagrangian. Since $\mathcal{O}^I(\varphi_\mathrm{B})$ is a basis, this is indeed the most general expression we can write. At the renormalized level, we rephrase the same expression as

$$-(L^I + h^I(L))\mathcal{O}_\mathrm{R}^I(\varphi), \qquad (2.107)$$

where the functions $h^I(L)$ collect the counterterms that are at least quadratic in the renormalized sources L^I. They subtract the divergences of the correlation functions that contain more than one insertion of $\mathcal{O}_\mathrm{R}^I(\varphi)$, such as the right one of (2.102).

The (bare and renormalized) generating functionals Z, W and Γ are defined as usual. Now, they depend on the sources L, besides J, or Φ. At the bare level, we have

$$Z_B(J_B, L_B) = \int [d\varphi_B] \exp\left(-S_B(\varphi_B, L_B) + \int J_B \varphi_B\right) = e^{W_B(J_B, L_B)},$$

$$(2.108)$$

where

$$S_B(\varphi_B, L_B) = S_B(\varphi_B) - \int L_B^I \mathcal{O}^I(\varphi_B) \qquad (2.109)$$

is the extended action. The composite field is minus the functional derivative of $S_B(\varphi_B, L_B)$ with respect to L_B^I. Its insertion into a (connected) correlation function is the functional derivative of W_B with respect to L_B^I.

Written in terms of renormalized quantities, the expressions (2.108) and (2.109) read

$$Z(J, L) = \int [d\varphi] \exp\left(-S(\varphi, L) + \int J\varphi\right) = e^{W(J, L)},$$

$$S(\varphi, L) = S(\varphi) - \int (L^I + h^I(L))\mathcal{O}_R^I(\varphi).$$

The correlation functions that carry $\mathcal{O}_R^I$ insertions can be obtained by differentiating the generating functionals with respect to L^I. The sources L are inert under the Legendre transform that switches from W to Γ.

Since $\mathcal{O}_B^I$ and $\mathcal{O}_R^J$ are bases of composite fields, there exists a matrix Z^{IJ} of renormalization constants, such that

$$\mathcal{O}_B^I = Z^{IJ} \mathcal{O}_R^J. \qquad (2.110)$$

We organize the $\mathcal{O}^I$'s into a column, such that the composite fields of equal dimensions are close to one another, and the composite fields of higher dimensions are placed below those of lower dimensions. Since the theory, by the renormalizability assumption, contains only parameters of nonnegative dimensions, a composite field can only mix with composite fields of equal or smaller dimensions. For this reason, the matrix Z_{IJ} is block lower triangular. Each diagonal block encodes the renormalization mixing of the composite fields of equal dimensions. The off-diagonal blocks encode the mixing among composite fields of different dimensions.

Comparing the L sectors of the bare and the renormalized actions, we find

$$-L_B^I \mathcal{O}_B^I = -(L^I + h^I(L))\mathcal{O}_R^I = -(L^I + h^I(L))(Z^{-1})^{IJ}\mathcal{O}_B^J, \qquad (2.111)$$

whence

$$L_B^I = (L^J + h^J(L))(Z^{-1})^{JI}. \qquad (2.112)$$

The structure of $\mathcal{O}_R^I$ as a vertex is more clearly visible when it is written in terms of renormalized (elementary) fields φ. Using formula (2.110), we have

$$\mathcal{O}_R^I = (Z^{-1})^{IJ}\mathcal{O}_B^J(\varphi_B) = (Z^{-1})^{IJ}\mathcal{O}_B^J(Z_\varphi^{1/2}\varphi),$$

which makes it easy to interpret $-(L^I + h^I(L))\mathcal{O}_R^I$ as a collection of vertices.

The correlation functions are studied diagrammatically, with the procedure we are already familiar with. The extended actions replace the original actions. The Feynman rules are supplemented with the vertices generated by (2.106), or (2.107). We use them to calculate, and renormalize, all the diagrams that can be built by including the new vertices. The source L^I are viewed as external, nonpropagating fields. In particular, the diagrams have no L internal legs.

For example, in the case of $\mathcal{O} = \varphi^2/2$ (including the standard combinatorial factor for the permutations of identical legs), the vertex to be added to the Lagrangian is $-L\varphi^2/2$, with one leg L and two legs φ. Once $L\mathcal{O}$ is written in terms of the renormalized fields, the new vertex is treated as any other vertex. The source L is an external field.

We add

$$-\frac{1}{2}\int \mathrm{d}^D x\, L_B(x)\varphi_B^2(x) = -\frac{Z_{\varphi^2}^{-1} Z_\varphi}{2}\int \mathrm{d}^D x\, L(x)\varphi^2(x) \qquad (2.113)$$

to the bare action (2.47), and

$$= 1 \qquad (2.114)$$

to the Feynman rules.

Observe that the source L has dimension two in units of mass, so power counting tells us that the action can include a counterterm that is quadratic in L and contains no field φ, which renormalizes the second correlation function of (2.102). We can view it as a renormalization of the source L_0 coupled with the identity operator. We write its bare and renormalized versions as

$$-\int L_{0B} = -\int L_0 - \frac{\delta_0 \mu^{-\varepsilon}}{2} \int L^2. \tag{2.115}$$

The generating functional becomes

$$Z(J, L) = e^{W(J,L)} = \int [d\varphi] \, \exp\left(-S(\varphi, L) + \int J\varphi\right), \tag{2.116}$$

where the extended action reads

$$S(\varphi, L) = S(\varphi) - \frac{Z_{\varphi^2}^{-1} Z_\varphi}{2} \int L\varphi^2 - \int L_0 - \frac{\delta_0 \mu^{-\varepsilon}}{2} \int L^2,$$

and $S(\varphi)$ is given by (2.16). Since L is an external field, the Feynman diagrams can have external legs L, but no internal legs of that type. There are only two overall divergent correlation functions that contain φ^2 insertions, which are precisely those of equation (2.102). The counterterms associated with them give Z_L and δ_0, which we now calculate at one loop.

The first correlation function is corrected by the diagram

$$\tag{2.117}$$

which is $(-\lambda \mu^\varepsilon/2)\mathcal{I}_D(k, m)$, where $\mathcal{I}_D$ is written in formula (2.29). We easily find

$$Z_{\varphi^2}^{-1} = 1 + \frac{\lambda}{16\pi^2 \varepsilon} + \mathcal{O}(\lambda^2).$$

The φ^2 two-point function is given at one loop by the diagram

$$\tag{2.118}$$

which is $(1/2)\mathcal{I}_D(k, m)$, so

$$\delta_0 = -\frac{1}{16\pi^2 \varepsilon} + \mathcal{O}(\lambda). \tag{2.119}$$

Exercise 17 *Calculate the functionals $W(J, L)$ and and its Legendre transform $\Gamma(\Phi, L)$ with respect to J for a free massless scalar field in the presence of the composite field φ^2.*

Solution. We have the renormalized generating functional

$$e^{W(J,L)} = \int [\mathrm{d}\varphi] \exp\left(-\frac{1}{2}\int \left[(\partial_\mu\varphi)^2 - L\varphi^2 - 2L_0 - \mu^{-\varepsilon}\delta_0 L^2\right] + \int J\varphi\right),$$
$$(2.120)$$

where δ_0 is given by (2.119) with $\lambda = 0$. The functional integral is easy to work out, since it is Gaussian. The source L plays the role of a spacetime dependent mass. We obtain

$$W(J, L) = \frac{1}{2}\int \left[J\frac{1}{-\Box - L}J + 2L_0 + \mu^{-\varepsilon}\delta_0 L^2\right] - \frac{1}{2}\mathrm{tr}\ln(-\Box - L),$$

$$\Gamma(\Phi, L) = \frac{1}{2}\int \left[(\partial_\mu\Phi)^2 - L\Phi^2 - 2L_0 - \mu^{-\varepsilon}\delta_0 L^2\right] + \frac{1}{2}\mathrm{tr}\ln(-\Box - L),$$
$$(2.121)$$

where

$$\Phi = \int \frac{1}{-\Box - L}J. \qquad (2.122)$$

□

Let us further comment on the multiple insertions of composite fields, i.e., the terms of $S(\varphi, L)$ that contain quadratic or higher powers of the sources L^I. In general, the renormalized action $S(\varphi, L)$ is not polynomial in L^I. Indeed, if the dimension of $\mathcal{O}_B^I$ is large, the dimension of L_B^I is negative. Then, infinitely many local counterterms with high powers of the sources L_B^I and their derivatives can be built. By the no-miracle principle of renormalization, $S(\varphi, L)$ must contain all of them. This means, in particular, that, strictly speaking, $S(\varphi, L)$ is not even local, since it contains terms with arbitrarily many derivatives. However, it is perturbatively local, since each order of the perturbative expansion is local. At any rate, we do not need to worry, because we are not required to resum the L^I powers. Every correlation function contains a given, finite number of composite-field insertions, so it can be calculated by truncating $S(\varphi, L)$ to appropriate finite powers of L^I. Every truncation is local and polynomial. Thus, we can still call $S(\varphi, L)$ a local functional, according to the extended definition of local functionals introduced before.

Exercise 18 *Calculate the one-loop renormalization of the composite field* $\mathcal{O}(\varphi) = \varphi^4$ *and the composite fields that mix with it, in the* φ^4 *theory.*

Solution. At one loop φ^4 mixes only with φ^2, while φ^2 mixes only with the identity operator. Add $-L_4\varphi^4/4! - L_2\varphi^2/2!$ to the classical Lagrangian, calculate the counterterms that renormalize it at one loop, then differentiate the renormalized Lagrangian with respect to the sources L_4 and L_2 to get (minus) the renormalized operators (equipped with the combinatorial factors). The results are

$$\frac{[\varphi^4]}{4!} = \left(1 + \frac{3\lambda}{8\pi^2\varepsilon}\right)\frac{\varphi^4}{4!} + \frac{m^2\mu^{-\varepsilon}}{16\pi^2\varepsilon}\frac{\varphi^2}{2}, \qquad \frac{[\varphi^2]}{2} = \left(1 + \frac{\lambda}{16\pi^2\varepsilon}\right)\frac{\varphi^2}{2} + \frac{m^2\mu^{-\varepsilon}}{16\pi^2\varepsilon}.$$

Recall that $\varphi_{\mathrm{B}} = \varphi$ to this order.

2.8 Maximum poles of diagrams

An L-loop diagram has at most poles $1/\varepsilon^L$ of order L. Sometimes the order of its maximum pole can be considerably smaller than L. For example, exercise 5 shows that the diagram (k) of figure (2.60) has two loops, but just a simple pole, at $m = 0$.

Here we prove a general theorem binding the maximum pole of a diagram. Since we are only interested in the UV divergences, and their renormalization, it is consistent to treat the mass terms, if present, as vertices with two legs. Then the propagator is just the massless one. Any other dimensionful parameter that multiplies a quadratic term must be treated in a similar way. To avoid IR problems in the intermediate steps, it is convenient to calculate the UV divergences of the Feynman diagrams by means of deformed propagators that are equipped with an artificial mass δ, and let δ tend to zero at the end, as explained in formula (2.81). The tadpoles are loops with a single vertex, and their divergent parts vanish identically (at $\delta = 0$). Instead, the loops with at least two vertices are not tadpoles (even if one of the vertices is a two-leg "mass" vertex) and may give nontrivial divergent contributions.

Theorem 3 *The maximum pole of a diagram with V vertices and L loops is at most*
$$\frac{1}{\varepsilon^{m(V,L)}},$$

where

$$m(V, L) = \min(V - 1, L).$$

Proof. We prove the statement inductively in V and, for fixed V, inductively in L. The diagrams with $V = 1$ and arbitrary L are tadpoles, which vanish identically and trivially satisfy the theorem. Suppose that the statement is true for $V < \bar{V}$, $\bar{V} > 1$, and arbitrary L. Consider the diagrams that have $\bar{V}$ vertices. For $L = 1$ the maximum pole is $1/\varepsilon$, so the theorem is satisfied. Proceed inductively in L, i.e. suppose that the theorem is also satisfied by the diagrams that have $\bar{V}$ vertices and $L < \bar{L}$ loops, and consider the diagrams $G_{\bar{V},\bar{L}}$ that have $\bar{V}$ vertices and $\bar{L}$ loops. If $G_{\bar{V},\bar{L}}$ has no subdivergence, its divergence is at most a simple pole, which satisfies the theorem. Higher-order poles are related to the subdivergences of $G_{\bar{V},\bar{L}}$, and can be classified by replacing the subdiagrams with their counterterms. Consider the subdiagrams $\gamma_{v,l}$ of $G_{\bar{V},\bar{L}}$ that have l loops and v vertices. Clearly, $1 \leqslant l < \bar{L}$ and $1 \leqslant v \leqslant \bar{V}$. By the inductive hypotheses, the maximum divergence of $\gamma_{v,l}$ is a pole of order $m(v, l)$. Contract the subdiagram $\gamma_{v,l}$ to a point and multiply by $1/\varepsilon^{m(v,l)}$. A diagram with $\bar{V} - v + 1 \leqslant \bar{V}$ vertices and $\bar{L} - l < \bar{L}$ loops is obtained, whose maximum divergence is at most a pole of order $m(v, l) + m(\bar{V} - v + 1, \bar{L} - l)$, if we take the factor $1/\varepsilon^{m(v,l)}$ into account. The inequality

$$m(v, l) + m(\bar{V} - v + 1, \bar{L} - l) \leqslant m(\bar{V}, \bar{L}),$$

which can be derived case by case, proves that the maximum divergence of $G_{\bar{V},\bar{L}}$ associated with $\gamma_{v,l}$ satisfies the theorem. Since this is true for every subdiagram $\gamma_{v,l}$, the theorem follows for $G_{\bar{V},\bar{L}}$. By induction, the theorem follows for every diagram. $\square$

Recall that this theorem holds after expanding in powers of the dimensionful parameters that are contained in the propagators. The diagram (k) of figure (2.60) has $V = 2$ and $L = 2$, so $m(V, L) = 1$: indeed, its maximum pole in the massless limit is a simple pole instead of a double pole. It can be easily checked that at $m \neq 0$ the diagram has a double pole proportional to the squared mass. If we view the mass term as a two-leg vertex, that pole arises from the diagram obtained from (k) by attaching the two-leg vertex to one internal line. In that case, we have $V = 3$ and $L = 2$, so $m(V, L) = 2$, in agreement with the theorem.

2.9   Subtraction prescription

When we subtract a simple pole $1/\varepsilon$, we can equivalently subtract an arbitrary finite constant together with it, as shown in formula (2.33). Similarly,
when we subtract a multiple pole $1/\varepsilon^n$, we can affect the less singular poles:

$$\frac{1}{\varepsilon^n} \to \frac{1}{\varepsilon^n} + \sum_{i=1}^{n} \frac{c_i}{\varepsilon^{n-i}}.$$

Sometimes, a prescription, called *subtraction scheme*, is adopted to associate
finite constants c_i to the subtractions of the poles, according to a convenient
rule. The *minimal* subtraction (MS) scheme is the convention according to
which the poles are subtracted with no finite constants attached.

By the locality of the counterterms, the scheme arbitrariness can only
affect the local terms. This means that it amounts to a finite redefinition of
the constants that multiply the vertices and the kinetic terms contained in
the Lagrangian. Since those constants, including the field normalizations,
are arbitrary anyway, the arbitrariness amounts to a finite reparametrization
of the theory, and does not affect the physical quantities.

In other words, renormalization is an infinite reparametrization of the
theory, while a change of subtraction scheme is a finite reparametrization.

To be more explicit, consider the vertex φ^4 and its one-loop counterterm
(2.33):

$$\lambda\mu^\varepsilon \frac{\varphi^4}{4!} + 3\lambda^2\mu^\varepsilon \left(\frac{1}{16\pi^2\varepsilon} + c_1 \right) \frac{\varphi^4}{4!}. \tag{2.123}$$

Now, move the arbitrary constant c_1 from the counterterm to the vertex φ^4
and define

$$\lambda'(\lambda) = \lambda + 3\lambda^2 c_1 + \mathcal{O}(\lambda^3). \tag{2.124}$$

We can rewrite (2.123) as

$$\mu^\varepsilon \lambda' \frac{\varphi^4}{4!} + \mu^\varepsilon \frac{3\lambda'^2}{16\pi^2\varepsilon} \frac{\varphi^4}{4!} + \mathcal{O}(\lambda'^3). \tag{2.125}$$

We see that the finite reparametrization (2.124) converts the arbitrary subtraction (2.123) to the minimal form (2.125). It is always possible to make
a similar rearrangement.

From the experimental point of view, the arbitrariness disappears when
enough physical quantities are measured. Specifically, in the massive φ^4

theory two independent quantities need to be measured. From them, the values of m and λ can be derived (once the φ normalization is established conventionally), after which the theory is uniquely determined. Observe that the parameter m needs not be identified with the physical mass, sometimes denoted with m_{ph}. Since m_{ph} can only be a finite function of m and λ, it is determined once m and λ are.

In the minimally subtracted λ' parametrization (2.125), the theory does not depend on c_1, so it is uniquely determined once λ' is measured (together with m, having decided the φ normalization). On the other hand, in the nonminimally subtracted λ parametrization (2.123) there appears to be an additional arbitrary constant c_1, so it seems that an extra measurement is necessary. This is just a blunder, because after the two measurements mentioned above and the φ normalization, c_1 disappears from all the physical quantities.

The matter can be better explained as follows. Consider some physical quantity. Write it as a function $f(\lambda)$ of λ in the first scheme and a function $f'(\lambda')$ of λ' in the second scheme. When we change the scheme, we do not just change λ, but also the form of the function f of λ. The two changes compensate each other, so that the physical results remain the same, that is to say,

$$f(\lambda) = f'(\lambda').$$

Check for example (2.123) and (2.125): the coupling changes, but also the function multiplying φ^4 changes, so that (2.123) and (2.125) coincide. So, if an experimental measurement gives $\lambda' = \ell$ in the second scheme, where ℓ is some number that we assume to be small, a measurement in the first scheme gives the number

$$\lambda = \ell - 3\ell^2 c_1 + \mathcal{O}(\ell^3),$$

whatever the value of c_1 is. If c_1 is left arbitrary, the measurement just gives the function $\lambda'(\lambda, c_1)$, but not λ and c_1 separately.

2.10   Regularization prescription

So far, we have mostly worked with the dimensional regularization, but equivalent results can be obtained with any regularization technique we like.

Indeed, switching from one technique to another one is equivalent to changing the subtraction scheme, and has no physical impact. To prove this statement, it is helpful to clarify the very definition of regularization technique.

Definition 3 *We define the formal limit as the limit in which the regularization parameters are removed by keeping the bare fields and bare parameters fixed.*

We emphasize that, in spite of its name, the formal limit is a rigorous notion. The formal limit of the action is the classical action. The formal limit of the Feynman rules is well defined. The formal limit of the correlation functions is in general ill defined, because of the divergences.

Definition 4 *We define the physical limit as the limit in which the regularization parameters are removed by keeping the renormalized fields and renormalized parameters fixed.*

The physical limit of the action is ill defined, but the physical limit of the correlation functions exists.

Consider a quantum field theory $\mathcal{T}$, defined by an action $S(\varphi)$ and a functional measure $[d\varphi]$.

Definition 5 *A regularized theory for $\mathcal{T}$ is a deformed theory $\mathcal{T}_{\mathcal{R}}$, defined by a deformed action $S_{\mathcal{R}}(\varphi)$ and a deformed functional measure $[d_{\mathcal{R}}\varphi]$, such that: (i) all the regularized diagrams are convergent; (ii) the propagators and the vertices tend to the ones of $\mathcal{T}$ in the formal limit; and (iii) all the diagrams, or derivatives of diagrams, that are convergent at the unregularized level are recovered by taking the formal limits of their regularized versions.*

Now, consider an integral $\int f$ and define two regularized versions of it,

$$\int f_1(\Lambda_1) < \infty, \qquad \int f_2(\Lambda_2) < \infty, \qquad (2.126)$$

Λ_1 and Λ_2 denoting regularization parameters. We just call them cutoffs and assume that they are removed by sending them to infinity. By definition, we must have

$$\lim_{\Lambda_1 \to \infty} f_1(\Lambda_1) = \lim_{\Lambda_2 \to \infty} f_2(\Lambda_2) = f. \qquad (2.127)$$

Indeed, the integrands contain vertices and propagators, namely, ingredients inherited from the classical action, so they must tend to f in the formal limits. However, we cannot extend the limits to the integrals, because the latter might be divergent. This is the reason why the limits are called formal.

Expanding for large $\Lambda_{1,2}$, we can write

$$\int f_i(\Lambda_i) = I_{i\text{div}}(\Lambda_i) + I_{i\text{finite}} + I_{i\text{ev}}(\Lambda_i), \qquad (2.128)$$

where $i = 1, 2$, while $I_{i\text{div}}$ collects the terms that diverge, $I_{i\text{ev}}$ those that tend to zero and $I_{i\text{finite}}$ those that have finite limits.

We know that (assuming that the subdivergences have been subtracted away by means of the usual algorithmic procedure), the integrals become convergent at the unregularized level, once we take a sufficient number of derivatives with respect to the external momenta k: there exists an n such that

$$\int \frac{\partial^n f}{\partial k^n} < \infty. \qquad (2.129)$$

This property is independent of the regularization technique.

Now, because of (2.127), we also have

$$\lim_{\Lambda_1 \to \infty} \frac{\partial^n}{\partial k^n} f_1(\Lambda_1) = \lim_{\Lambda_2 \to \infty} \frac{\partial^n}{\partial k^n} f_2(\Lambda_2) = \frac{\partial^n f}{\partial k^n}.$$

Integrating each side of this equation, and using (2.129), we obtain

$$\int \lim_{\Lambda_1 \to \infty} \frac{\partial^n}{\partial k^n} f_1(\Lambda_1) - \int \lim_{\Lambda_2 \to \infty} \frac{\partial^n}{\partial k^n} f_2(\Lambda_2) = 0.$$

The regularization conditions (2.126) ensure that the integrals are convergent even before taking the derivatives and the limits, so we can move the integration signs right before the functions f_i. This gives

$$\lim_{\Lambda_1 \to \infty} \frac{\partial^n}{\partial k^n} \int f_1(\Lambda_1) - \lim_{\Lambda_2 \to \infty} \frac{\partial^n}{\partial k^n} \int f_2(\Lambda_2) = 0.$$

Using (2.128) and singling out the Λ_i-dependent and the Λ_i-independent sectors, we get

$$\frac{\partial^n}{\partial k^n} I_{1\text{div}}(\Lambda_1) = \frac{\partial^n}{\partial k^n} I_{2\text{div}}(\Lambda_2) = 0, \qquad \frac{\partial^n}{\partial k^n} \left(I_{1\text{finite}} - I_{2\text{finite}} \right) = 0.$$

The first formula follows from the fact that the two derivatives must be equal, so they can neither depend on Λ_1, nor Λ_2. Since they must depend on them (because they are divergent), they are zero. The result contained in the first formula is just the statement that the counterterms are local with any regularization technique. The second formula, instead, states that the finite parts, calculated using two different regularization techniques, can differ at most by local terms, i.e., a polynomial in k:

$$I_{1\text{finite}} = I_{2\text{finite}} + \text{local}.$$

If the theory is renormalizable, such local terms are of the types already present in the Lagrangian, so they amount to a scheme change and do not affect the physical quantities. This concludes the proof.

Sometimes it is convenient to regularize different classes of diagrams in different ways, or introduce multiple cutoffs Λ_i. The divergences expressed in terms of different cutoffs can be identified up to local terms. The cutoffs Λ_i can be removed in different orders, e.g., $\Lambda_1 \to \infty$ followed by $\Lambda_2 \to \infty$, or $\Lambda_2 \to \infty$ followed by $\Lambda_1 \to \infty$. When the limits are interchanged, the results can differ at most by local terms, i.e., again a scheme change, but the physical quantities are always the same.

Ultimately, we have an enormous freedom. We can regularize a theory as a whole, or diagram by diagram. We can use one cutoff or many cutoffs, and we can remove the cutoffs in the order we like. We can even use a different regularization technique and a different subtraction scheme for each diagram. No matter how we regularize the theory, the physical results always come out right. The core of quantum field theory is finite and regularization independent. The divergences are confined to the "superficial" parts of the integrals, so to speak, since they are killed by a finite number of derivatives. For this reason, they can be removed by means of a set of "superficial" operations.

Every result holds in nonrenormalizable theories as well, with the sole difference that we must introduce infinitely independent couplings, and organize the perturbative expansion in combination with suitable truncations in the powers of the parameters that have negative dimensions.

2.11   Comments about the dimensional regularization

Some people use to say that the dimensional regularization "misses something" or "has problems of internal consistency", because integrals such as (2.12) are set to zero and the powerlike divergences disappear, or because of other caveats that we have not met, yet.

The truth is that the dimensional regularization does not miss anything, and has no problems of internal consistency. Actually, it is the most powerful regularization technique developed so far. It is convenient both to make calculations (to the extent that the renormalization of QCD has been worked out to four loops, and the one of the standard model to three loops) and to prove theorems to all orders. No alternative regularization technique is even comparable, in these respects, with the dimensional one.

One of the virtues of the dimensional regularization is that it automatically chooses a subtraction scheme where the powerlike divergences are absent. Actually, it allows us to prove that, no matter what regularization technique we use, the powerlike divergences are devoid of any physical meaning. They can always be subtracted away just as they come, without leaving any remnants. Later we will better understand what this means, studying the renormalization group. For the moment, it is sufficient to note that the powerlike divergences are scheme dependent, which is why they can be removed completely by means of a smart choice of subtraction scheme. Instead, the logarithmic divergences are only partially scheme dependent, which is why they cannot be removed by changing the subtraction scheme, and do contain physical information.

Two attitudes may lead to skepticism towards the dimensional regularization. One is the assumption that behind empty space there should be a sort of lattice, or "aether". The analytic way to regularize the integrals is not intuitive, some people say, while a lattice spacing is supposed to be more "physical". Probably, those people should also explain why a regularization should be intuitive, or "physical", and what that is supposed to mean. We just observe that sometimes analogies with condensed matter physics or other domains may be inspiring, but other times they just put us on the wrong track. More generally, it sounds gratuitous to assume that high-energy physics should conform to some other branch of physics. It is also contrary to the trend exhibited by data in more than a century. Not only that, but there is no reason why we should assume that human intuition (which is always the product of our interaction with a classical environment) is a good guide. It may be helpful in some cases, misleading in others. Once we have given up the correspondence principle almost completely, we can live without intuition, or content ourselves with a partial intuition.

Another attitude is the assumption that the ultimate theory of the universe should be finite, that is to say, a theory with no divergences. In that case, the powerlike divergences are not really divergences, but physical quantities that depend on a large energy scale, and grow polynomially with it. The assumption that the final theory should be finite turns out to be "appealing" to some people (for reasons that are more emotional than physical). The point is that it is rather restrictive. Having learned that we can renormalize the divergences away, we no longer need to require that they are absent from the start. If one insists that the final theory must be finite, he or she should explain why, given that we can make sense of theories that are not finite, we should privilege a small subset of the theories we can make sense of, and ignore the others. Somebody explicitly advocates aesthetic criteria to "answer" these questions. Other people try to disguise their subjective requirements under suspicious conditions of "simplicity". Certainly, simplicity can be advocated for practical purposes, but it cannot be advocated to discriminate what is physical from what is not: that part pertains to nature. Letting aside that literally *anything* can become simple *a posteriori*, once one gets used to it. At any rate, we do not think it is necessary to stress further that the oblique arguments we have reported so far in this section, as popular as they may be, are completely meaningless in physics.

Another point about the dimensional regularization is that it is "just" perturbative. However, at present we do not know how to define the functional integral nonperturbatively, so this problem goes beyond the dimensional regularization itself.

2.12 About the series resummation

We have stressed several times that our task is to define the functional integral as a perturbative expansion. We have converted the functional integral of an interacting theory into an infinite sum of functional integrals of a free field theory, since those are the only ones we can deal with. Each property we state, or use, must be understood in the same spirit. For example, when we say that the integral of a functional total derivative vanishes, we mean that each term of the perturbative expansion vanishes. At this level, the perturbative expansion must be regarded as a sequence, a list of terms, not as a series that should be resummed.

Our primary objective is to define the terms of the sequence, and check that they are consistent with the key physical and mathematical requirements of quantum field theory. As we have seen, this task already raises nontrivial problems. Several other difficulties will appear in gauge theories and dealing with anomalies. It makes no sense to investigate the resummation properties before defining the terms of the sequence.

In various cases, the sum of the perturbative series might not even exist, at least naïvely. This might mean that different ways to organize the sum can give different results. Then, we must classify the resummation prescriptions that make physical sense. There might be more meaningful resummation prescriptions, each of them leading to a different physical theory, with the same perturbative expansion. Recalling that, so far, we have not been scared away by the divergences (and now we appreciate what we would have missed, if we had), there is no point in worrying about a problem that is not even there, yet.

We will actually see that often we have control on the perturbative expansion to arbitrarily high orders (such as in the cases of the particle widths, the anomalies, the renormalization-group flow and the conformal fixed points). In all those cases the series does make sense, or the theory itself provides a natural resummation prescription for it. For example, there are anomalies that can be calculated exactly, since they receive no corrections beyond one loop (if we are careful enough, which includes choosing an appropriate subtraction scheme).

In this book, we make no attempt to define the functional integral beyond its perturbative expansion, unless that means searching for the physical prescriptions that allow us to resum the perturbative expansion when possible.

Chapter 2 — References

[1] The analytic regularization was first introduced in 1964 by Bollini, Giambiagi and Gonzáles Domínguez, in the paper

C.G. Bollini, J.J. Giambiagi and A. Gonzáles Domínguez, *Analytic regularization and the divergences of quantum field theories*, Nuovo Cim. 31 (1964) 550.

[2] Elaborating on those ideas, Bollini and Giambiagi came up with the dimensional regularization in 1971:

C.G. Bollini and J.J. Giambiagi, *The number of dimensions as a regularizing parameter*, Nuovo Cim. 12B (1972) 20;

C.G. Bollini and J.J. Giambiagi, *Lowest order divergent graphs in ν-dimensional space*, Phys. Lett. B40, 566 (1972).

[3] Right away, 't Hooft and Veltman elaborated on it in great detail:

G. 't Hooft and M. Veltman, *Regularization and renormalization of gauge fields*, Nucl. Phys. B 44 (1972) 189.

[4] A classic paper that contains mathematical aspects of the dimensional regularization is

P. Breitenlohner and D. Maison, *Analytic renormalization and the action principle*, Commun. Math. Phys. 52 (1977) 11

[5] The dimensional regularization of chiral theories has various annoying features. Several modifications have been studied in the literature. A deformation that has the virtue of simplifying the proofs of properties to all orders can be found in

D. Anselmi, *Weighted power counting and chiral dimensional regularization*, Phys. Rev. D 89 (2014) 125024, 14A2 Renorm and arXiv:1405.3110 [hep-th]

[6] The theorem of section 2.8 about the maximum poles of diagrams first appeared in D. Anselmi, *Renormalization of a class of nonrenormalizable theories*, J. High Energ. Phys. 07 (2005) 077, 05A1 Renorm and arXiv:hep-th/0502237 [hep-th].

Chapter 3

Renormalization group

In this chapter we begin to study the physical consequences of renormalization. Our considerations are very general, although we often illustrate them by means of specific models. We start by comparing the bare and the renormalized actions

$$S_B(\varphi_B, \lambda_B, m_B^2, L_B) = S_R(\varphi, \lambda, m^2, L, \mu) \qquad (3.1)$$

of a theory of fields φ, where λ and m denote the dimensionless and the dimensionful parameters, respectively, and L are the sources coupled to the composite fields. We do not really need to keep λ and m distinct, but for the moment it is convenient to do so. Similarly, the relation between the bare and the renormalized Γ functionals is

$$\Gamma_B(\Phi_B, \lambda_B, m_B^2, L_B) = \Gamma_R(\Phi, \lambda, m^2, L, \mu). \qquad (3.2)$$

We recall that in a theory with a single field φ and a single coupling λ, such as the φ_4^4 theory, we have relations of the form

$$\varphi_B = Z_\varphi^{1/2}(\lambda, \varepsilon)\varphi, \qquad \lambda_B = \mu^{\sigma\varepsilon}\lambda Z_\lambda(\lambda, \varepsilon),$$
$$m_B^2 = m^2 Z_{m^2}(\lambda, \varepsilon), \qquad L_B^I = \tilde{L}^J(\lambda, \varepsilon)(Z^{-1}(\lambda, \varepsilon))^{JI}, \qquad (3.3)$$

for some coefficient σ, where we have introduced the (divergent) combinations $\tilde{L}^I(\lambda, \varepsilon) = L^I + h^I(L)$, for convenience, from formula (2.111). The theories with more couplings and fields have more complicated relations, which we discuss later. The key point is that the renormalized sides of (3.1) and (3.2) depend on one quantity more than the bare sides. Precisely,

137

the renormalized sides depend on λ and μ separately, while the bare sides contain only λ_B, which is a specific combination of λ and μ. Therefore, it must be possible to solve the μ dependence exactly in terms of the renormalization constants. The solution can be obtained by comparing the bare correlation functions to the renormalized ones. Their μ dependence is called renormalization-group (RG) flow.

Let us see what originates the dependence on μ. At the tree level, the action depends on a unique combination of λ and μ, such as $\lambda\mu^{\sigma\varepsilon}$ in the case (3.3). However, that combination cannot survive beyond the tree level, because the subtraction of divergences is an operation that separates μ from λ: the counterterms are multiplied by higher powers of λ, which is dimensionless, but by the same power of μ, which is fixed by the dimensional analysis (see for example (2.34)). This gives the final relation $\lambda_B = \mu^{\sigma\varepsilon}\lambda Z_\lambda(\lambda, \varepsilon)$.

The key quantities that are used to describe the renormalization-group flow are the beta function and the anomalous dimensions. To introduce them, we define the total derivative $\mu\mathrm{d}/\mathrm{d}\mu$, which is the derivative calculated by keeping the bare quantities fixed, and the partial derivative $\mu\partial/\partial\mu$, which is the derivative calculated by keeping the renormalized quantities fixed. When we apply the total derivative to the functional Γ we obtain, by the Leibniz rule,

$$
\mu\frac{\mathrm{d}}{\mathrm{d}\mu} = \mu\frac{\partial}{\partial\mu} + \mu\frac{\mathrm{d}\lambda}{\mathrm{d}\mu}\frac{\partial}{\partial\lambda} + \mu\frac{\mathrm{d}m^2}{\mathrm{d}\mu}\frac{\partial}{\partial m^2} + \int \mathrm{d}^D x\, \mu\frac{\mathrm{d}\Phi(x)}{\mathrm{d}\mu}\frac{\delta}{\delta\Phi(x)}
$$
$$
+ \int \mathrm{d}^D x\, \mu\frac{\mathrm{d}\tilde{L}^I(x)}{\mathrm{d}\mu}\frac{\delta}{\delta\tilde{L}^I(x)}. \tag{3.4}
$$

If we apply the total derivative to the action, we obtain the same formula with Φ replaced by φ. If we apply the total derivative to the functional W, we obtain the same formula with Φ replaced by J.

Clearly, the action and the generating functionals have vanishing total derivatives with respect to μ, because they are μ independent when written in terms of bare quantities:

$$
\mu\frac{\mathrm{d}S_B}{\mathrm{d}\mu} = \mu\frac{\mathrm{d}S_R}{\mathrm{d}\mu} = \mu\frac{\mathrm{d}W_B}{\mathrm{d}\mu} = \mu\frac{\mathrm{d}W_R}{\mathrm{d}\mu} = \mu\frac{\mathrm{d}\Gamma_B}{\mathrm{d}\mu} = \mu\frac{\mathrm{d}\Gamma_R}{\mathrm{d}\mu} = 0. \tag{3.5}
$$

The functional integration measure shares the same property, for the same reason.

Beta function

Define the "hat beta function" as

$$\hat{\beta}_\lambda = \mu \frac{d\lambda}{d\mu}.$$

At the tree level, $\lambda_B = \mu^{\sigma\varepsilon}\lambda$, so $\hat{\beta}_\lambda = -\sigma\varepsilon\lambda + \mathcal{O}(\lambda^2)$. It is convenient to define the *beta function* β_λ so that

$$\hat{\beta}_\lambda(\lambda, \varepsilon) = \beta_\lambda(\lambda, \varepsilon) - \sigma\varepsilon\lambda. \tag{3.6}$$

Clearly, $\beta_\lambda = \mathcal{O}(\lambda^2)$. If we apply the identity (3.4) to λ_B, and recall that Z_λ depends only on λ and ε, we find

$$0 = \mu \frac{d\lambda_B}{d\mu} = \left(\mu \frac{\partial}{\partial\mu} + \hat{\beta}_\lambda \frac{\partial}{\partial\lambda}\right)(\mu^{\sigma\varepsilon}\lambda Z_\lambda) = \sigma\varepsilon\mu^\varepsilon\lambda Z_\lambda + \mu^\varepsilon \hat{\beta}_\lambda \frac{d(\lambda Z_\lambda)}{d\lambda},$$

whence

$$\frac{d\ln Z_\lambda}{d\lambda} = -\frac{\beta_\lambda}{\lambda\hat{\beta}_\lambda}. \tag{3.7}$$

Using (3.6), we infer

$$\beta_\lambda = \frac{\sigma\varepsilon\lambda^2 \frac{d\ln Z_\lambda}{d\lambda}}{1 + \lambda\frac{d\ln Z_\lambda}{d\lambda}}. \tag{3.8}$$

We also derive the inverse formula

$$Z_\lambda(\lambda, \varepsilon) = \exp\left(-\int_0^\lambda \frac{d\lambda'}{\lambda'} \frac{\beta_\lambda(\lambda', \varepsilon)}{\beta_\lambda(\lambda', \varepsilon) - \sigma\varepsilon\lambda'}\right). \tag{3.9}$$

The lower integration limit is fixed by demanding $Z_\lambda(0, \varepsilon) = 1$, since the renormalization constants of free fields are equal to one.

Anomalous dimension

Let us study the total derivative of φ_B. Using (3.4) with $\Phi \to \varphi$ we find

$$0 = \mu \frac{d\varphi_B}{d\mu} = \mu \frac{d}{d\mu}\left(Z_\varphi^{1/2}\varphi\right) = \mu \frac{dZ_\varphi^{1/2}}{d\mu}\varphi + Z_\varphi^{1/2}\mu \frac{d\varphi}{d\mu},$$

that is to say,

$$\mu \frac{d\varphi}{d\mu} = -\gamma_\varphi \varphi,$$

where

$$\gamma_\varphi \equiv \frac{1}{2}\mu\frac{d\ln Z_\varphi}{d\mu} = \frac{1}{2}\mu\frac{d\lambda}{d\mu}\frac{d\ln Z_\varphi}{d\lambda} = \frac{1}{2}\hat\beta_\lambda\frac{d\ln Z_\varphi}{d\lambda}. \tag{3.10}$$

The quantity γ_φ is called *anomalous dimension* of the field φ, and depends on λ and ε. Since $\Phi = \langle\varphi\rangle$, and the action, as well as the functional integration measure, has vanishing total derivative with respect to μ, we also obtain

$$\mu\frac{d\Phi}{d\mu} = -\gamma_\varphi\Phi.$$

From (3.10) we find the inverse formula

$$Z_\varphi(\lambda,\varepsilon) = \exp\left(2\int_0^\lambda d\lambda'\frac{\gamma_\varphi(\lambda',\varepsilon)}{\beta_\lambda(\lambda',\varepsilon) - \sigma\varepsilon\lambda'}\right).$$

When the renormalization is multiplicative, we have

$$\int J_B\varphi_B = \int J\varphi,$$

so the renormalization constants of φ and J are the reciprocals of each other. Then we also find

$$\mu\frac{dJ}{d\mu} = \gamma_\varphi J. \tag{3.11}$$

Exercise 19 *Calculate the beta functions and the anomalous dimensions of the φ_4^4 theory and the φ_6^3 theory at one loop in the minimal subtraction scheme.*

Solution. In the minimal subtraction scheme, the constants c_1 and c_2 of formula (2.45) are equal to zero. Applying the formulas (3.8) and (3.10) to (2.45), (2.48) and (2.49), we get

$$\varphi_4^4: \quad \beta_\lambda = \frac{3\lambda^2}{16\pi^2} + \mathcal{O}(\lambda^3), \qquad \gamma_\varphi = \mathcal{O}(\lambda^2), \tag{3.12}$$

$$\varphi_6^3: \quad \beta_\lambda = -\frac{3\lambda^3}{4(4\pi)^3} + \mathcal{O}(\lambda^5), \qquad \gamma_\varphi = \frac{\lambda^2}{12(4\pi)^3} + \mathcal{O}(\lambda^4), \tag{3.13}$$

Exercise 20 *Calculate the beta function and the anomalous dimension of the φ_4^4 theory at two loops.*

Solution. Applying the formulas (3.8) and (3.10) to (2.65) and (2.66) we get

$$\beta_\lambda = \frac{3\lambda^2}{(4\pi)^2} - \frac{17\lambda^3}{3(4\pi)^4} + \mathcal{O}(\lambda^4), \qquad \gamma_\varphi = \frac{\lambda^2}{12(4\pi)^4} + \mathcal{O}(\lambda^3).$$

Note that the divergences have canceled out. Later we will prove that this is a general fact.

Exercise 21 *Calculate first nonvanishing contributions to the beta function and the anomalous dimension of the massless φ_3^6 theory.*

Solution. Applying the formulas (3.8) and (3.10) to (2.67) and (2.69), and recalling that here $\sigma = 2$, we get

$$\beta_\lambda = \frac{5\lambda^2}{3(4\pi)^2} + \mathcal{O}(\lambda^3), \qquad \gamma_\varphi = \frac{\lambda^2}{90(8\pi)^4} + \mathcal{O}(\lambda^3). \qquad (3.14)$$

Composite fields

We start from the correlation functions that contain a single insertion of some composite field $\mathcal{O}^I$, plus insertions of elementary fields. This means that we can neglect the terms that are quadratic in L in (3.4) and (2.111), which gives

$$L_B^I \mathcal{O}_B^I = L^I \mathcal{O}_R^I = L^I (Z^{-1})^{IJ} \mathcal{O}_B^J \equiv L^I Z_L^{IJ} \mathcal{O}_B^J,$$

where $Z_L^{IJ} = (Z^{-1})^{IJ}$ can be seen as the renormalization constants of the sources L, so that $L_B^I = L^J Z_L^{JI}$. Let us also assume, for the moment, that we are considering an operator $\mathcal{O}$ that has no renormalization mixing with other operators. By exercise 18, examples are the operators φ^4 and φ^2 in the massless φ_4^4 theory, at one loop. Taking the total derivative $\mu d/d\mu$ of L_B, we get

$$0 = \mu \frac{dL_B}{d\mu} = L\mu \frac{dZ_L}{d\mu} + \mu \frac{dL}{d\mu} Z_L,$$

whence

$$\mu \frac{dL}{d\mu} = -\mu \frac{d\ln Z_L}{d\mu} L = -\hat{\beta}_\lambda \frac{d\ln Z_L}{d\lambda} L.$$

From $L_B \mathcal{O}_B = L\mathcal{O}$, $Z_L = Z_{\mathcal{O}}^{-1}$, we find

$$\mu \frac{d\mathcal{O}}{d\mu} = -\gamma_{\mathcal{O}} \mathcal{O}, \qquad \gamma_{\mathcal{O}} = -\gamma_L = \mu \frac{d\ln Z_{\mathcal{O}}}{d\mu} = -\mu \frac{d\ln Z_L}{d\mu}.$$

The inverse formula reads

$$Z_{\mathcal{O}}(\lambda, \varepsilon) = \exp\left(\int_0^{\lambda} d\lambda' \frac{\gamma_{\mathcal{O}}(\lambda', \varepsilon)}{\beta_{\lambda}(\lambda', \varepsilon) - \sigma \varepsilon \lambda'} \right). \tag{3.15}$$

Consider the mass operator $\varphi^2/2$ in four dimensions. Its renormalization coincides with the renormalization of the integrated mass term

$$\int d^D x \frac{\varphi^2}{2}. \tag{3.16}$$

Indeed, the integral determines the integrand up to local total derivatives, which in this case must have dimension 2. Since there exist no local Lorentz invariant object with these features, φ^2 and (3.16) renormalize in exactly the same way. Correspondingly, the source L_{φ^2} coupled to φ^2 renormalizes as m^2, and $Z_{\varphi^2} = Z_{m^2}^{-1}$. By dimensional analysis, $\mu dm^2/d\mu$ must be equal to m^2 times a function of λ and ε. We have

$$\mu \frac{dm^2}{d\mu} = m^2 \eta(\lambda, \varepsilon), \qquad \eta = -\mu \frac{d \ln Z_{m^2}}{d\mu} = \mu \frac{d \ln Z_{\varphi^2}}{d\mu} = \gamma_{\varphi^2}.$$

More generally, the composite fields mix with one another. As explained in section 2.7, it is convenient to collect all of them into a huge vector $\mathcal{O}^I$, where the composite fields of the same dimension are close to one another, and the composite fields of higher dimensions follow those of lower dimensions. Then, we have formula (2.110), where the matrix Z^{IJ} of renormalization constants is block lower triangular. We find

$$\mu \frac{d[\mathcal{O}^I]}{d\mu} = -\gamma_{IJ}[\mathcal{O}^J], \qquad \gamma_{IJ} \equiv Z_{IK}^{-1} \mu \frac{dZ^{KJ}}{d\mu}. \tag{3.17}$$

Finally, when we want to include the multiple insertions of composite fields, we can replace the sources L^I with the expressions $\tilde{L}^I$. The sources $\tilde{L}^I$ coupled to the renormalized operators $\mathcal{O}^I$ satisfy

$$\mu \frac{d\tilde{L}^I}{d\mu} = \tilde{L}^J \gamma_{JI}. \tag{3.18}$$

3.1   The Callan-Symanzik equation

Let us apply (3.4) to $\Gamma(\Phi)$ (no composite fields). We obtain

$$0 = \mu \frac{d\Gamma_{\mathrm{B}}}{d\mu} = \mu \frac{\partial \Gamma}{\partial \mu} + \hat{\beta}_{\lambda} \frac{\partial \Gamma}{\partial \lambda} + \eta m^2 \frac{\partial \Gamma}{\partial m^2} - \gamma_{\varphi} \int d^D x\, \Phi(x) \frac{\delta \Gamma}{\delta \Phi(x)}. \tag{3.19}$$

On W we have, instead,

$$0 = \mu \frac{\mathrm{d}W_{\mathrm{B}}}{\mathrm{d}\mu} = \mu \frac{\partial W}{\partial \mu} + \hat{\beta}_\lambda \frac{\partial W}{\partial \lambda} + \eta m^2 \frac{\partial W}{\partial m^2} + \gamma_\varphi \int \mathrm{d}^D x \, J(x) \frac{\delta W}{\delta J(x)}. \tag{3.20}$$

Let us take two functional derivatives of (3.19) with respect to Φ and set $\Phi = 0$ afterwards, or, equivalently, two derivatives of (3.20) with respect to J and set $J = 0$ afterwards. We obtain the Callan-Symanzik equations for the irreducible two-point function Γ_2 and the connected two-point function $W_2 = \langle \varphi(x)\varphi(y) \rangle_c$:

$$0 = \left(\mu \frac{\partial}{\partial \mu} + \hat{\beta}_\lambda \frac{\partial}{\partial \lambda} + \eta m^2 \frac{\partial}{\partial m^2} - 2\gamma_\varphi \right) \Gamma_2, \tag{3.21}$$

$$0 = \left(\mu \frac{\partial}{\partial \mu} + \hat{\beta}_\lambda \frac{\partial}{\partial \lambda} + \eta m^2 \frac{\partial}{\partial m^2} + 2\gamma_\varphi \right) W_2. \tag{3.22}$$

The two equations are indeed equivalent, because $\Gamma_2 W_2 = 1$.

For the moment, let us work with the massless theory. We do not make the ε dependence explicit every time, because it is not important for the present discussion. Since W_2 has dimension $D - 2$ it is convenient to write

$$W_2(|x - y|, \lambda, \mu) = \frac{G_{2r}(t, \lambda)}{|x - y|^{D-2}}, \qquad t \equiv -\ln(|x - y|\mu). \tag{3.23}$$

Then (3.22) at $m = 0$ becomes

$$0 = \left(-\frac{\partial}{\partial t} + \hat{\beta}_\lambda \frac{\partial}{\partial \lambda} + 2\gamma_\varphi \right) G_{2r}(t, \lambda). \tag{3.24}$$

We want to solve this equation. To this purpose, we define the "running coupling" $\tilde{\lambda}(t, \lambda)$, which is the solution of the first-order differential problem

$$\frac{\mathrm{d}\tilde{\lambda}}{\mathrm{d}t} = \hat{\beta}_\lambda(\tilde{\lambda}), \qquad \tilde{\lambda}(0, \lambda) = \lambda. \tag{3.25}$$

We have $\mathrm{d}t = \mathrm{d}\tilde{\lambda}/\hat{\beta}_\lambda(\tilde{\lambda})$, hence

$$t = \int_\lambda^{\tilde{\lambda}(t,\lambda)} \frac{\mathrm{d}\tilde{\lambda}}{\hat{\beta}_\lambda(\tilde{\lambda})}. \tag{3.26}$$

It is convenient to view $\tilde{\lambda}$ as a function of both t and the initial condition λ. If so, the t derivative appearing in (3.25) must be written as a partial derivative $\partial\tilde{\lambda}(t,\lambda)/\partial t$. Differentiating each side of (3.26) with respect to λ, we can work out the derivative of the solution with respect to its initial condition, which is

$$\frac{\partial\tilde{\lambda}(t,\lambda)}{\partial\lambda} = \frac{\hat{\beta}_\lambda(\tilde{\lambda}(t,\lambda))}{\hat{\beta}_\lambda(\lambda)}. \tag{3.27}$$

The solution of the Callan-Symanzik equation (3.24) reads

$$G_{2r}(t,\lambda) = z^{-1}(\lambda,t)G_{2r}(0,\tilde{\lambda}(t,\lambda)), \tag{3.28}$$

with

$$z(\lambda,t) = \exp\left(-2\int_0^t \gamma_\varphi(\tilde{\lambda}(s,\lambda))\mathrm{d}s\right). \tag{3.29}$$

We prove this statement by checking that (3.28) satisfies the equation and the initial condition. Given a function f of many variables, we write $f^{(n_1,n_2,\dots)}$ to denote its n_1th partial derivative with respect to its first variable, n_2th partial derivative with respect to its second variable, and so on.

The initial condition is certainly satisfied, since $t = 0$ gives the identity $G_{2r}(0,\lambda) = G_{2r}(0,\lambda)$. Moreover, we can easily calculate the partial derivatives of G_{2r} with respect to t and λ. We find

$$G_{2r}^{(1,0)}(t,\lambda) = 2\gamma_\varphi(\tilde{\lambda}(t,\lambda))G_{2r}(t,\lambda) + z^{-1}(\lambda,t)\hat{\beta}_\lambda(\tilde{\lambda}(t,\lambda))G_{2r}^{(0,1)}(0,\tilde{\lambda}(t,\lambda)),$$

$$G_{2r}^{(0,1)}(t,\lambda) = 2G_{2r}(t,\lambda)\int_0^t \frac{\partial\tilde{\lambda}(s,\lambda)}{\partial\lambda}\gamma_\varphi'(\tilde{\lambda}(s,\lambda))\mathrm{d}s$$

$$+z^{-1}(\lambda,t)\frac{\partial\tilde{\lambda}(t,\lambda)}{\partial\lambda}G_{2r}^{(0,1)}(0,\tilde{\lambda}(t,\lambda)).$$

Now, using (3.27) we also have

$$\hat{\beta}_\lambda(\lambda)\int_0^t \frac{\partial\tilde{\lambda}(s,\lambda)}{\partial\lambda}\gamma_\varphi'(\tilde{\lambda}(s,\lambda))\mathrm{d}s = \int_0^t \hat{\beta}_\lambda(\tilde{\lambda}(s,\lambda))\gamma_\varphi'(\tilde{\lambda}(s,\lambda))\mathrm{d}s \tag{3.30}$$

$$= \int_0^t \frac{\partial\tilde{\lambda}(s,\lambda)}{\partial s}\gamma_\varphi'(\tilde{\lambda}(s,\lambda))\mathrm{d}s = \int_0^t \frac{\partial\gamma_\varphi(\tilde{\lambda}(s,\lambda))}{\partial s}\mathrm{d}s = \gamma_\varphi(\tilde{\lambda}(t,\lambda)) - \gamma_\varphi(\lambda).$$

Summing, we find that (3.24) is satisfied.

When the theory contains more parameters λ_i (which can include also the masses), equation (3.24) becomes

$$0 = \left(-\frac{\partial}{\partial t} + \hat{\beta}_\lambda^i\frac{\partial}{\partial\lambda_i} + 2\gamma_\varphi\right)G_{2r}(t,\lambda). \tag{3.31}$$

Define the running parameters $\tilde{\lambda}_i(t, \lambda)$ as the solutions of the following system of first-order differential equations and initial conditions:

$$\frac{d\tilde{\lambda}_i}{dt} = \hat{\beta}^i_\lambda(\tilde{\lambda}), \qquad \tilde{\lambda}_i(0, \lambda) = \lambda_i. \tag{3.32}$$

The solution (3.28) and formula (3.29) remain the same. However, formulas (3.26) and (3.27) do not hold. Define

$$f_i(t, \lambda) \equiv \hat{\beta}^j_\lambda(\lambda) \frac{\partial \tilde{\lambda}_i(t, \lambda)}{\partial \lambda_j},$$

where the sum over j is understood. We have $f_i(0, \lambda) = \hat{\beta}^i_\lambda(\lambda)$. Moreover, if $\hat{\beta}^i_{\lambda,k}(\lambda) \equiv \partial \hat{\beta}^i_\lambda(\lambda) / \partial \lambda_k$, we get

$$\frac{\partial f_i}{\partial t} = \hat{\beta}^j_\lambda(\lambda) \frac{\partial \hat{\beta}^i_\lambda(\tilde{\lambda}(t, \lambda))}{\partial \lambda_j} = \hat{\beta}^j_\lambda(\lambda) \frac{\partial \tilde{\lambda}_k(t, \lambda)}{\partial \lambda_j} \hat{\beta}^i_{\lambda,k}(\tilde{\lambda}(t, \lambda)) = f_k \hat{\beta}^i_{\lambda,k}(\tilde{\lambda}(t, \lambda)).$$

We obtain the system of first-order equations and initial conditions

$$\frac{\partial f_i(t, \lambda)}{\partial t} = f_k(t, \lambda) \hat{\beta}^i_{\lambda,k}(\tilde{\lambda}(t, \lambda)), \qquad f_i(0, \lambda) = \hat{\beta}^i_\lambda(\lambda).$$

It is easy to check that

$$F_i(t, \lambda) \equiv \hat{\beta}^i_\lambda(\tilde{\lambda}(t, \lambda))$$

satisfies the equations and the initial conditions. In particular,

$$\frac{\partial F_i(t, \lambda)}{\partial t} = \hat{\beta}^k_\lambda(\tilde{\lambda}(t, \lambda)) \hat{\beta}^i_{\lambda,k}(\tilde{\lambda}(t, \lambda)) = F_k(t, \lambda) \hat{\beta}^i_{\lambda,k}(\tilde{\lambda}(t, \lambda)).$$

We conclude that the equality $F_i(t, \lambda) = f_i(t, \lambda)$ holds, that is to say,

$$\hat{\beta}^j_\lambda(\lambda) \frac{\partial \tilde{\lambda}_i(t, \lambda)}{\partial \lambda_j} = \hat{\beta}^i_\lambda(\tilde{\lambda}(t, \lambda)). \tag{3.33}$$

This formula is a generalization of (3.27). Going through the proof of (3.28), we realize that (3.27) was necessary only to derive (3.30). Extending the proof of (3.28) to the theories that contain more parameters, we see that (3.33) is just sufficient to derive the desired generalization of (3.30).

In the end, we find that (3.28) satisfies (3.31), as wanted.

General solution of the Callan-Symanzik equation

So far, we have studied the two-point function. However, the results can be extended to a generic correlation function

$$W^{I_1\cdots I_m}_{\alpha_1\cdots\alpha_n}(x,y,\lambda,\mu) = \langle \varphi_{\alpha_1}(x_1)\cdots\varphi_{\alpha_n}(x_n)\mathcal{O}^{I_1}(y_1)\cdots\mathcal{O}^{I_m}(y_m)\rangle_c \quad (3.34)$$

that contains both insertions of elementary and composite fields. The subscript α in φ_α is used to distinguish different types of elementary fields, while λ collects the couplings and every other parameter, including the masses. We denote the φ_α anomalous dimensions with $\gamma_\alpha(\lambda)$.

The Callan-Symanzik equation for (3.34) can be derived by applying (3.4) (with $\Phi \to J$) to $W(J,L)$ and using (3.11) and (3.18). We find

$$0 = \left(\mu\frac{\partial}{\partial\mu} + \sum_i \hat{\beta}^i_\lambda \frac{\partial}{\partial\lambda_i} + \sum_{i=1}^n \gamma_{\alpha_i}\right) W^{I_1\cdots I_m}_{\alpha_1\cdots\alpha_n}$$
$$+ \sum_{j=1}^m \gamma_{I_j K_j} W^{I_1\cdots I_{j-1}K_j I_{j+1}\cdots I_m}_{\alpha_1\cdots\alpha_n} + \mathrm{CT}_L, \quad (3.35)$$

where CT_L stands for "contact terms" for the composite fields. They appear because the insertion of a renormalized composite field is a functional derivative with respect to L^I (not $\tilde{L}^I$, which would give a divergent result). Using (3.18), the last term of (3.4) gives on W

$$\int \mu\frac{\mathrm{d}\tilde{L}^I}{\mathrm{d}\mu}\frac{\delta W}{\delta\tilde{L}^I} = \int \tilde{L}^J\gamma_{JI}\frac{\delta W}{\delta\tilde{L}^I} = \int \tilde{L}^J\gamma_{JI}\frac{\delta L^K}{\delta\tilde{L}^I}\frac{\delta W}{\delta L^K} = \int L^J\tilde{\gamma}_{JI}(L)\frac{\delta W}{\delta L^I},$$

where

$$\tilde{\gamma}_{JI}(L) \equiv (\delta^{JM} + h^{JM}(L))\gamma_{MK}\frac{\delta L^I}{\delta\tilde{L}^K} = \gamma_{JI} + \ L\text{-dependent corrections},$$

and $h^{JM}(L)$ is the matrix such that $L^J h^{JM}(L) = h^M(L)$ (where the sum on the repeated index J may understand an integral over spacetime, and partial integrations may be necessary to factorize L^J). Now, any time we differentiate with respect to L, we have to differentiate $\tilde{\gamma}_{JI}(L)$ as well, which generates corrections proportional to (possibly, derivatives of) a delta function $\delta^{(D)}(y_i - y_j)$ in (3.35). Thus, the contact terms CT_L of (3.35) are distributions of this type, times correlation functions with fewer insertions. They can be studied similarly to what we are doing right now for the main contributions to (3.35). There is no loss of information if we ignore these types of contact terms for the moment (by assuming that $y_i \neq y_j$ for every $i \neq j$), and resume them along the lines outlined here when needed.

Repeating the proof of the previous section, it is easy to show that the solution reads

$$W^{I_1 \cdots I_m}_{\alpha_1 \cdots \alpha_n}(x, y, \lambda, \mu) = \prod_{i=1}^n z_{\alpha_i}^{-1/2}(t) \prod_{j=1}^m Z^{-1}_{I_j K_j}(t) \, W^{K_1 \cdots K_m}_{\alpha_1 \cdots \alpha_n}(x, y, \tilde{\lambda}(t), \tilde{\mu}) + \mathrm{CT}_L,$$
(3.36)

where now $t = \ln(\tilde{\mu}/\mu)$. The running couplings $\tilde{\lambda}(t)$ are the solutions of (3.32), as before. Moreover,

$$z_{\alpha_i}(t) = \exp\left(-2\int_0^t \mathrm{d}s\gamma_{\alpha_i}(s)\right), \quad Z^{-1}(t) = T\exp\left(\int_0^t \mathrm{d}s\gamma(s)\right), \quad (3.37)$$

where Z and γ are the matrices with entries Z_{IJ} and γ_{IJ}, respectively, and $\gamma(t)$ stands for $\gamma(\tilde{\lambda}(t))$. Finally,

$$Z^{-1}(t) = 1 + \sum_{k=1}^\infty \int_0^t \mathrm{d}t_1 \int_0^{t_1} \mathrm{d}t_2 \cdots \int_0^{t_{k-1}} \mathrm{d}t_k \gamma(t_1) \cdots \gamma(t_{k-1})\gamma(t_k). \quad (3.38)$$

That is to say, T denotes the "time"-ordered product for $t > 0$, and the "time"-anti-ordered product (combined with a sign flip of γ) for $t < 0$.

Thus, formula (3.36) tells us how the correlation function depends on the scale μ.

3.2   Finiteness of the beta function and the anomalous dimensions

Formulas (3.12) and (3.54) show that the poles in ε cancel out, at one-loop, in the beta functions, as well as in the anomalous dimensions. This is a very general fact: the beta functions and the anomalous dimensions are convergent quantities. Consider the Callan-Symanzik equation (3.19) for $\Gamma(\Phi, \lambda, m^2, \mu)$. Restore $\hbar$ and expand each quantity perturbatively,

$$\Gamma = \sum_{i=0}^\infty \hbar^i \Gamma_i, \qquad \hat{\beta}_\lambda = \sum_{i=0}^\infty \hbar^i \hat{\beta}_{\lambda i}, \qquad \eta = \sum_{i=0}^\infty \hbar^i \eta_i, \qquad \gamma_\varphi = \sum_{i=0}^\infty \hbar^i \gamma_{\varphi i}.$$
(3.39)

Assume inductively that $\hat{\beta}_\lambda$, η and γ_φ are convergent up to and including the order $n-1$, that is to say

$$\hat{\beta}_{\lambda i}, \eta_i, \gamma_{\varphi i} < \infty \qquad \text{for } i \leqslant n-1. \tag{3.40}$$

The assumption is obviously true for $i = 0$, since $\hat{\beta}_{\lambda 0} = -\varepsilon\lambda$, $\eta_0 = \gamma_{\varphi 0} = 0$. Consider the contribution to (3.19) of order n. We have

$$0 = \mu\frac{\partial\Gamma_n}{\partial\mu} + \sum_{i=0}^{n}\left(\hat{\beta}_{\lambda i}\frac{\partial\Gamma_{n-i}}{\partial\lambda} + \eta_i m^2\frac{\partial\Gamma_{n-i}}{\partial m^2} - \gamma_{\varphi i}\int\Phi\frac{\delta\Gamma_{n-i}}{\delta\Phi}\right).$$

Recall that every Γ_i is convergent, and so are its derivatives with respect to the renormalized parameters. Using (3.40) we conclude

$$\hat{\beta}_{\lambda n}\frac{\partial\Gamma_0}{\partial\lambda} + \eta_n m^2\frac{\partial\Gamma_0}{\partial m^2} - \gamma_{\varphi n}\int\Phi\frac{\delta\Gamma_0}{\delta\Phi} = \text{finite}. \tag{3.41}$$

Now, Γ_0 is just the classical action, and $\partial\Gamma_0/\partial\lambda$, $\partial\Gamma_0/\partial m^2$ and $\Phi(\delta\Gamma_0/\delta\Phi)$ are independent terms, because they are the vertex, the mass term and the field equation (which contains the vertex, the mass term and the kinetic term), respectively. Therefore, each coefficient of the linear combination (3.41) must be separately finite, which proves

$$\hat{\beta}_{\lambda n} < \infty, \qquad \eta_n < \infty, \qquad \gamma_{\varphi n} < \infty.$$

The inductive assumption (3.40) is thus promoted to $n = \infty$.

We have set $L^I = 0$, but the argument can be repeated in the presence of sources for the composite fields $\mathcal{O}^I$. This proves that the quantities γ_{IJ} are also finite, as long as the $\mathcal{O}^I$'s are independent from one another.

3.3   Fixed points of the RG flow

Consider the correlation function $W_{\alpha_1\cdots\alpha_n}^{I_i\cdots I_m}$ of formula (3.34), with $y_i \neq y_j$ for every $i \neq j$. We rescale the coordinates, the momenta and the parameters λ by powers of ζ equal to their dimensions in units of mass. For example, the coordinates x^μ are rescaled to $\zeta^{-1}x^\mu$, the momenta p^μ to ζp^μ, the masses m to ζm, and so on. If we also rescale μ to $\zeta\mu$, we get

$$W_{\alpha_1\cdots\alpha_n}^{I_i\cdots I_m}(\zeta^{-1}x, \zeta^{-1}y, \zeta^{d_\lambda}\lambda, \zeta\mu) = \zeta^{d_W}W_{\alpha_1\cdots\alpha_n}^{I_i\cdots I_m}(x, y, \lambda, \mu),$$

where d_W and d_λ are the dimensions of $W^{I_i \cdots I_m}_{\alpha_1 \cdots \alpha_n}$ and λ, respectively. Replacing μ by $\zeta^{-1}\mu$, we obtain

$$W^{I_i \cdots I_m}_{\alpha_1 \cdots \alpha_n}(x, y, \lambda, \zeta^{-1}\mu) = \zeta^{-d_W} W^{I_i \cdots I_m}_{\alpha_1 \cdots \alpha_n}(\zeta^{-1}x, \zeta^{-1}y, \zeta^{d_\lambda}\lambda, \mu).$$

The left-hand side of this equation tells us that the limit $\mu \to \infty$ in the correlation function is equivalent to letting ζ tend to zero. The right-hand side tells us that this operation is equivalent to letting the distances tend to infinity (and rescale the parameters of the theory appropriately). Thus, the limit $\mu \to \infty$ gives information about the infrared, or large-distance, behaviors of the correlation functions. Similarly, the limit $\mu \to 0$ is equivalent to taking ζ to infinity and, in particular, the distances to zero, so it gives information about the ultraviolet behaviors.

The solution (3.36) of the renormalization-group equations (with $\mu \to \zeta^{-1}\mu$, $\tilde{\mu} \to \mu$) gives

$$W^{I_i \cdots I_m}_{\alpha_1 \cdots \alpha_n}(x, y, \lambda, \zeta^{-1}\mu) = \prod_{i=1}^{n} z_{\alpha_i}^{-1/2}(t) \prod_{j=1}^{m} Z^{-1}_{I_j K_j}(t) \, W^{K_1 \cdots K_m}_{\alpha_1 \cdots \alpha_n}(x, y, \tilde{\lambda}(t), \mu),$$

(3.42)

where now $t = \ln \zeta$. Thus, to understand the infrared and the ultraviolet behaviors of the correlation functions, it is useful to work out the infrared and the ultraviolet behaviors of the beta functions and the anomalous dimensions.

For simplicity, we assume that the theory has a unique dimensionless coupling, and keep calling it λ. We also assume that λ is defined to be nonnegative. Typically, as in the case of the theory φ^4, this requirement is necessary to ensure that the potential is bounded from below.

An alternative way of defining the running coupling constant is by viewing it as a function $\lambda(\mu)$ of the energy scale μ. Start from formula (3.26), and set $\varepsilon = 0$, $t = \ln(\tilde{\mu}/\mu)$, $\tilde{\lambda} = \lambda(\tilde{\mu})$ and $\lambda = \lambda(\mu)$. In a generic subtraction scheme, define $\beta_\lambda(\lambda) \equiv \hat{\beta}_\lambda(\lambda, 0)$. Exponentiating (3.26), relabeling the integration variable, and splitting the integral into two symmetric parts with the help of an arbitrary constant $\bar{\lambda}$, we can write

$$\tilde{\mu} \exp\left(-\int_{\bar{\lambda}}^{\lambda(\tilde{\mu})} \frac{d\lambda}{\beta_\lambda(\lambda)}\right) = \mu \exp\left(-\int_{\bar{\lambda}}^{\lambda(\mu)} \frac{d\lambda}{\beta_\lambda(\lambda)}\right) = \text{constant} \equiv \Lambda_{\rm T}.$$

The scale Λ_T (called Λ_{QCD} if the theory is QCD and $\bar{\lambda}$ is chosen appropriately) is RG invariant, i.e., independent of μ. We also have

$$\int_{\bar{\lambda}}^{\lambda(\mu)} \frac{d\lambda}{\beta_\lambda(\lambda)} = \ln \frac{\mu}{\Lambda_T} \qquad (3.43)$$

Now, the infrared (ultraviolet) behavior of $\tilde{\lambda}$ is studied by taking $t \to -\infty$ ($t \to \infty$), which is equivalent to taking the limit $\tilde{\mu} \to 0$ ($\tilde{\mu} \to \infty$) of the function $\tilde{\lambda} = \lambda(\tilde{\mu})$. Then, it is also the limit $\mu \to 0$ ($\mu \to \infty$) of $\lambda(\mu)$. In both the infrared and the ultraviolet limits, the right-hand side of (3.43) diverges. On the other hand, the left-hand side can diverge in the following two cases: (i) the running coupling tends to a zero of the beta function, i.e.,

$$\lim_{\mu \to 0} \lambda(\mu) = \lambda_{IR}, \quad \beta_\lambda(\lambda_{IR}) = 0, \qquad \text{and/or}$$
$$\lim_{\mu \to \infty} \lambda(\mu) = \lambda_{UV}, \quad \beta_\lambda(\lambda_{UV}) = 0;$$

or (ii) the running coupling tends to $\pm\infty$. In all such cases the integral of (3.43) must diverge appropriately, to match the right-hand side.

The values of the couplings where the beta functions vanish at $\varepsilon = 0$, i.e., the solutions λ_* of $\beta_\lambda(\lambda_*) = 0$, define a particular class of quantum field theories, which are called fixed points of the RG flow. Clearly, $\lambda = 0$ is a trivial example of a fixed point, and corresponds to the free theory we have been expanding around. However, there may exist interacting fixed points, that is to say, solutions with $\lambda_* \neq 0$. This happens, for example, when the beta function has the forms

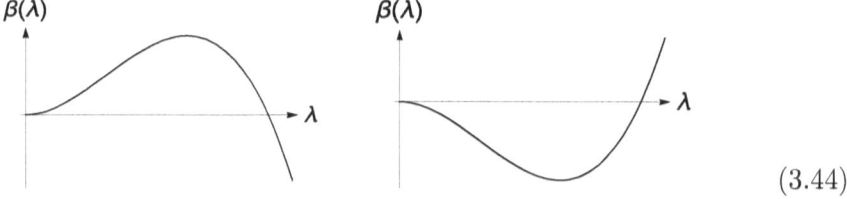

$$(3.44)$$

In some cases, the nontrivial fixed point can be reached perturbatively from the free one. Then, the perturbative expansion allows us to follow the entire renormalization-group flow in between the fixed points.

In a fixed point, the coupling does not run (at $\varepsilon = 0$), since $\beta_\lambda \equiv 0$ implies $\tilde{\lambda} \equiv \lambda_*$. However, a theory behaves rather differently around free and the interacting fixed points. Now we study the typical behaviors.

Expand β_λ perturbatively around $\lambda = 0$:

$$\beta_\lambda(\lambda) = \beta_1 \lambda^2 + \beta_2 \lambda^3 + \beta_3 \lambda^4 + \mathcal{O}(\lambda^5). \tag{3.45}$$

If the running coupling $\tilde{\lambda}$ is small, we can keep the first nontrivial contribution $\beta_1 \lambda^2$ in the RG equation (3.25), and neglect the higher orders. The running coupling then reads

$$\tilde{\lambda}(t, \lambda) = \frac{\lambda}{1 - \beta_1 t \lambda} \tag{3.46}$$

up to higher-order corrections. Setting $t = \ln(\tilde{\mu}/\mu)$, $\tilde{\lambda} = \lambda(\tilde{\mu})$ and $\lambda = \lambda(\mu)$, we can also express the running in the form (3.43), or

$$\frac{1}{\lambda(\mu)} + \beta_1 \ln \mu = \text{constant}. \tag{3.47}$$

However, this result is only valid to the lowest nontrivial order, and can be trusted only if the running coupling is small. This happens for $\mu \to 0$ (which is the IR limit) if $\beta_1 > 0$[1], and for $\mu \to \infty$ (which is the UV limit) if $\beta_1 < 0$.

We can derive the same conclusion from formula (3.46), which gives

$$\tilde{\lambda}(t, \lambda) \sim -\frac{1}{\beta_1 t} \ll 1 \qquad \text{for } |t| \gg 1. \tag{3.48}$$

The running coupling tends to zero (and the theory tends to the free fixed point) for $t \to +\infty$ if $\beta_1 < 0$, for $t \to -\infty$ if $\beta_1 > 0$. Observe that the behavior (3.48) is λ independent.

The theory is said to be *infrared free* if the running coupling tends to zero in the infrared limit ($\beta_1 > 0$), and *asymptotically free* (or ultraviolet free) if the running coupling tends to zero in the ultraviolet limit ($\beta_1 < 0$).

Examples of infrared-free theories are the φ_4^4 theory and the φ_3^6 theory, as shown by formulas (3.12) and (3.14), as well as quantum electrodynamics (see Chapter 6). Examples of asymptotically free theories are: (*i*) the φ_6^3 theory (check formula (3.13) and switch to the coupling $\alpha = \lambda^2$, whose beta function is $\beta_\alpha = 2\lambda\beta_\lambda$); (*ii*) the two-dimensional four-fermion model (1.106); and (*iii*) non-Abelian gauge field theories (Chapter 7), which make the strong interactions weak at high energies. In the case (*ii*), the one-loop beta function

$$\beta_\lambda = -\frac{(N-1)\lambda^2}{2\pi} + \mathcal{O}(\lambda^3) \tag{3.49}$$

---

[1]Recall that λ is assumed to be non-negative, so we must study $\lambda \to 0^+$ in (3.47).

can be worked out from formula (2.99).

If $\beta_1 = 0$, we have to consider the first nontrivial contribution to the beta function. Let us assume it is the one proportional to λ^{n+1} for $n \geqslant 1$:

$$\beta_\lambda(\lambda) = \beta_n \lambda^{n+1} + \mathcal{O}(\lambda^{n+2}). \tag{3.50}$$

Then we can work on the coupling $\alpha \equiv \lambda^n$, whose beta function is $\beta_\alpha = n\beta_n \alpha^2 + \mathcal{O}(\alpha^{2+\frac{1}{n}})$. To the lowest order, we reach the same conclusions as before with $\tilde{\lambda} \to \tilde{\alpha}$, $\lambda \to \alpha$, $\beta_1 \to n\beta_n$.

Now we study the behavior of the theory around an interacting fixed point. We write $\lambda = \lambda_* + \eta$, and take $\eta \ll 1$. The Taylor expansion of the beta function gives, to the lowest order,

$$\beta_\lambda(\lambda) = \beta'_\lambda(\lambda_*)\eta + \mathcal{O}(\eta^2).$$

We assume that the slope $\beta'_\lambda(\lambda_*)$ of the beta function at the fixed point is nonvanishing. If not, we have to go to a higher order of the Taylor expansion, which takes us back to the behavior (3.50). The RG problem (3.25) becomes

$$\frac{\mathrm{d}\tilde{\eta}}{\mathrm{d}t} = \beta'_\lambda(\lambda_*)\tilde{\eta} + \mathcal{O}(\tilde{\eta}^2), \qquad \tilde{\eta}(0) = \eta,$$

so the running coupling reads

$$\tilde{\eta}(\tilde{\mu}) = \eta e^{\beta'_\lambda(\lambda_*)t}.$$

Writing $t = \ln(\tilde{\mu}/\mu)$, to switch to the form (3.43), we obtain

$$\eta(\mu)\mu^{-\beta'_\lambda(\lambda_*)} = \text{constant}. \tag{3.51}$$

With the help of this formula, we can reach the fixed point. We know that $\eta(\mu)$ must tend to zero there. This happens when $\mu \to \infty$ for $\beta'_\lambda(\lambda_*) < 0$ and when $\mu \to 0$ for $\beta'_\lambda(\lambda_*) > 0$. We learn that if the slope of the beta function is negative (positive) at the interacting fixed point, the fixed point is reached in the ultraviolet (infrared) limit.

These conclusions are also visible from the figures (3.44): in the first (second) example of (3.44) the theory is free in the infrared (ultraviolet) limit, and tends to the interacting fixed point in the ultraviolet (infrared) limit.

In an interacting fixed point, the anomalous dimensions $\gamma^*_{\alpha_i} \equiv \gamma_{\alpha_i}(\lambda_*)$, $\gamma^*_{\varphi^2} \equiv \gamma_{\varphi^2}(\lambda_*)$ and $\gamma^*_{IJ} \equiv \gamma_{IJ}(\lambda_*)$ are just constants. Then, formulas (3.37) with $t = \ln(\tilde{\mu}/\mu)$ give

$$z_{\alpha_i}(t) = \left(\frac{\mu}{\tilde{\mu}}\right)^{2\gamma^*_{\alpha_i}}, \quad Z(t) = \left(\frac{\mu}{\tilde{\mu}}\right)^{\gamma^*}.$$

Finally, formula (3.36) gives

$$W^{I_i \cdots I_m}_{\alpha_1 \cdots \alpha_n}(x, y, \lambda_*, \mu) = \left(\frac{\tilde{\mu}}{\mu}\right)^{\sum_{i=1}^n \gamma^*_{\alpha_i}} \prod_{j=1}^m \left(\frac{\tilde{\mu}}{\mu}\right)^{\gamma^*}_{I_j K_j} W^{K_1 \cdots K_m}_{\alpha_1 \cdots \alpha_n}(x, y, \lambda_*, \tilde{\mu}).$$

$$(3.52)$$

In the particular case of the two-point function $W_2 = \langle \varphi(x)\varphi(y)\rangle_c$ of a massless theory, formulas (3.23), (3.28) and (3.29) give, in $D = 4$,

$$\langle \varphi(x)\varphi(y)\rangle_c = \frac{\mu^{-2\gamma^*_\varphi} C_\varphi}{|x-y|^{2(1+\gamma^*_\varphi)}},$$

where C_φ is a constant. If we compare this result with the two-point function of the free fixed point, which is

$$\langle \varphi(x)\varphi(y)\rangle_c = \frac{1}{4\pi^2 |x-y|^2},$$

we see that the exponent is modified to twice the "critical exponent"

$$1 + \gamma^*_\varphi.$$

In turn, this is the sum of the naïve φ dimension, which is equal to one, plus γ^*_φ. Similarly, the two-point function of a composite field $\mathcal{O}$ of naïve dimension $d_\mathcal{O}$ is

$$\langle \mathcal{O}(x)\mathcal{O}(y)\rangle_c = \frac{\mu^{-2\gamma^*_\mathcal{O}} C_\mathcal{O}}{|x-y|^{2(d_\mathcal{O}+\gamma^*_\mathcal{O})}}.$$

These remarks justify the name "anomalous dimensions" for the quantities γ.

The solution (3.52) is simple at the fixed points, because the Callan-Symanzik equation (3.35) becomes simpler there. Dropping the terms proportional to the beta functions, we get

$$0 = \left(\mu\frac{\partial}{\partial\mu} + \sum_{i=1}^n \gamma_{\alpha_i}\right) W^{I_1 \cdots I_m}_{\alpha_1 \cdots \alpha_n} + \sum_{j=1}^m \gamma_{I_j K_j} W^{I_1 \cdots I_{j-1} K_j I_{j+1} \cdots I_m}_{\alpha_1 \cdots \alpha_n},$$

so the entire μ dependence of a correlation function is encoded in an appropriate multiplicative factor.

φ_4^4 **at one loop** The RG flow of the theory φ^4 in four dimensions can be worked out at one loop by means of the beta function (3.12). Formula (3.12) is also correct at $\varepsilon \neq 0$ in the minimal subtraction scheme, while it contains corrections of the form $\mathcal{O}(\varepsilon\lambda^2)$ in a generic scheme, where the constants c_1 and c_2 of formula (2.45) can be nonzero. In either case, the solution of (3.25) at $\varepsilon = 0$ reads

$$\tilde{\lambda}(t, \lambda) = \frac{\lambda}{1 - \frac{3t\lambda}{16\pi^2}}, \tag{3.53}$$

which is the running coupling in the one-loop approximation. Since $\beta_1 > 0$, the theory is infrared free.

The anomalous dimension of the composite operator φ^2 is

$$\eta = \gamma_{\varphi^2} = -\mu \frac{d \ln Z_{m^2}}{d\mu} = -\hat{\beta}_\lambda \frac{d \ln Z_{m^2}}{d\lambda} = \frac{\lambda}{16\pi^2} + \mathcal{O}(\lambda^2). \tag{3.54}$$

Let us study the φ^2 two-point function in the massless case. From (3.37) we have

$$z_{\varphi^2}(\lambda, t) = \exp\left(-\int_0^t \gamma_{\varphi^2}(\tilde{\lambda}(s, \lambda)) ds\right) = \left(1 - \frac{3t\lambda}{16\pi^2}\right)^{1/3}.$$

Applying the RG solution (3.36) with $\tilde{\mu} = |x - y|^{-1}$, we get (again at $\varepsilon = 0$)

$$\langle \varphi^2(x)\varphi^2(y)\rangle_c \equiv \frac{G_{\varphi^2 r}^{(2)}(t, \lambda)}{|x - y|^4} \sim \left(\frac{\lambda_f(x - y, \mu)}{\lambda}\right)^{2/3} \frac{G_{\varphi^2 r}^{(2)}(0, \lambda_f(x - y, \mu))}{|x - y|^4} \tag{3.55}$$

at large distances, where

$$\lambda_f(x - y, \mu) \equiv \frac{16\pi^2}{3 \ln(\mu|x - y|)}.$$

We cannot define a critical exponent here, since λ_f has a logarithmic behavior. The reason is that the slope β'_λ of the beta function vanishes in the free fixed point.

3.4 Scheme (in)dependence

Now we work out other useful properties of the beta function. Observe that, in the minimal subtraction scheme, the λ renormalization constant (which we denote by means of a bar) has the form $\bar{Z}_\lambda = 1 + \text{poles in } \varepsilon$. Thus, formula (3.8) gives

$$\bar{\beta}_\lambda = \varepsilon \times \text{poles} = \text{finite} + \text{poles}, \tag{3.56}$$

with no orders ε^n, $n > 0$. However, we just proved that $\bar{\beta}_\lambda$ is finite, so the poles that appear on the right-hand side of (3.56) must cancel out. This means that $\bar{\beta}_\lambda$ depends only on λ and not on ε: in the minimal subtraction scheme we can write

$$\widehat{\bar{\beta}}_\lambda(\lambda, \varepsilon) = \bar{\beta}_\lambda(\lambda) - \sigma\varepsilon\lambda.$$

We know that the coefficients of the poles $1/\varepsilon$ are scheme independent at one loop. For this reason, the one-loop coefficients of the beta functions and the anomalous dimensions are always scheme independent, at $\varepsilon = 0$.

Moreover, if a theory contains a unique dimensionless coupling λ, we can easily show that the one-loop and the two-loop coefficients of the beta function are both scheme independent at $\varepsilon = 0$. To see this, expand $\beta_\lambda(\lambda)$ as in (3.45). A scheme change amounts to a perturbative reparametrization of λ, which we write as

$$\lambda = \lambda' + a_2\lambda'^2 + a_3\lambda'^3 + \mathcal{O}(\lambda'^4). \tag{3.57}$$

We have

$$\beta'_\lambda(\lambda') = \mu\frac{d\lambda'}{d\mu} = \mu\frac{d\lambda}{d\mu}\frac{d\lambda'}{d\lambda} = \beta_\lambda(\lambda(\lambda'))\left(\frac{d\lambda}{d\lambda'}\right)^{-1} =$$
$$= \beta_1\lambda'^2 + \beta_2\lambda'^3 + \left(\beta_3 + a_2\beta_2 + (a_2^2 - a_3)\beta_1\right)\lambda'^4 + \mathcal{O}(\lambda'^5). \tag{3.58}$$

We see that the first two coefficients, and only those, are scheme independent. The result does not extend to $\varepsilon \neq 0$, since in that case we have to include reparametrizations of the form $\lambda = c(\varepsilon)\lambda' + \mathcal{O}(\lambda'^2)$, with $c(\varepsilon) = 1 + \mathcal{O}(\varepsilon)$.

With a suitable choice of a_2 and a_3, the third coefficient can be set to zero, for example

$$a_2 = 0, \qquad a_3 = \frac{\beta_3}{\beta_1}. \tag{3.59}$$

It is easy to prove that, with a suitable choice of the function $\lambda(\lambda')$, all the coefficients but the first two can be set to zero. However, this fact is nothing more than a curiosity. The two-loop beta function cannot be trusted as an exact formula, not even within the perturbative expansion. A warning that there is a problem somewhere is the β_1 in the denominator of (3.59). Typically, β_1 is linear in the number N of fields that circulate in the loops. Check for example formula (3.49). Nowhere can the perturbative expansion generate inverse powers of N.

Moreover, if the reparametrization (3.57) is too general, it can introduce spurious singularities at finite values of λ. For example, factors such as

$$\frac{\lambda^2}{\beta_\lambda(\lambda)} = \frac{1}{\beta_1} + \mathcal{O}(\lambda) \qquad (3.60)$$

could be easily generated. If we ignore the β_1 in the denominator (maybe because we are working with a given number of fields, and are not aware of the importance of the point raised above), such functions appear to have a perfectly good perturbative expansion around the free-fixed point of the RG flow. Nevertheless, they do not have a good behavior around an interacting fixed point, because of the denominator $\beta_\lambda(\lambda)$. If we make reparametrizations that involve expressions such as (3.60), we may lose the possibility of smoothly interpolating between two fixed points of the RG flow.

Finally, the "curiosity" mentioned above does not extend to the theories that contain more than one coupling. When we generalize the argument outlined in (3.57) and (3.58), both λ and β_λ become vectors, while the coefficients β_i become tensors with $i+2$ indices, and the coefficients a_i become tensors with $i+1$ indices. The coefficients of the transformed beta function can be set to zero by solving linear recursive equations that have $i+2$ indices, but their unknowns just have $i+1$ indices. The solution does not exist, in general.

3.5   A deeper look into the renormalization group

If we insert the one-loop values (2.45) of the φ^4 renormalization constants into the inverse formulas (3.9) and (3.15), we can reconstruct the renormalization constants Z_λ and, for example, Z_{φ^2}. Then we find something interesting. Working in the minimal subtraction scheme, we obtain

$$\bar{Z}_\lambda(\lambda, \varepsilon) = \frac{1}{1 - \frac{3\lambda}{16\pi^2\varepsilon}}, \qquad \bar{Z}_{\varphi^2}(\lambda, \varepsilon) = \left(1 - \frac{3\lambda}{16\pi^2\varepsilon}\right)^{1/3}. \qquad (3.61)$$

These results give the correct values (2.45) (at $c_1 = c_2 = 0$) to the first order in λ. Moreover, the two-loop double poles of $\bar{Z}_\lambda$ agree with those of formula

(2.66).

The expressions (3.61) tell us much more than this. To uncover their true content, it is convenient to discuss the effects of the higher-order corrections. With no loss of generality, we can take $\sigma = 1$, since the case $\sigma \neq 1$ can be obtained from $\sigma = 1$ by means of the replacement $\beta_\lambda \to \beta_\lambda/\sigma$. We insert them in (3.9) by writing the beta function as $\bar{\beta}_\lambda(\lambda) = \lambda \sum_{i=1}^\infty \bar{\beta}_i \lambda^i$. When we do so, we can organize the expansion as

$$\bar{Z}_\lambda(\lambda, \varepsilon) = \exp\left(\int_0^\lambda \frac{\mathrm{d}\lambda'}{\lambda'} \frac{\sum_{i=1}^\infty \bar{\beta}_i \frac{\lambda'^i}{\varepsilon}}{1 - \sum_{j=1}^\infty \bar{\beta}_j \frac{\lambda'^j}{\varepsilon}} \right)$$

$$= 1 + \bar{\beta}_1 \frac{\lambda}{\varepsilon} f_1\left(\bar{\beta}_1 \frac{\lambda}{\varepsilon}\right) + \sum_{i=2}^\infty \bar{\beta}_i \frac{\lambda^i}{\varepsilon} f_i\left(\bar{\beta}_1 \frac{\lambda}{\varepsilon}, \bar{\beta}_2 \frac{\lambda^2}{\varepsilon}, \cdots, \bar{\beta}_i \frac{\lambda^i}{\varepsilon}\right), \quad (3.62)$$

where the functions f_i are power series of their arguments, and receive contributions from the j-th orders, $j \leqslant i$, of the beta function. We see that the maximum poles λ^n/ε^n of Z_λ, $n > 0$ (even those that are originated by diagrams with arbitrarily many loops), are not affected by the higher-order corrections to the beta function: they depend only on the one-loop coefficient β_1. Resumming them, we find

$$\bar{Z}_\lambda(\lambda, \varepsilon) = \frac{1}{1 - \frac{\beta_1 \lambda}{\varepsilon}} + \sum_{i=2}^\infty \frac{\lambda^i}{\varepsilon} f_i.$$

The first two coefficients of the beta function contribute to the poles that have the form $(\lambda^2/\varepsilon)^n(\lambda/\varepsilon)^m$, with $n > 0$ and $m \geqslant 0$. However, they do not determine all of them, because the same powers of λ and ε can be obtained in different ways. For example, λ^4/ε^2 can be viewed as $(\lambda^2/\varepsilon)^2$, or $(\lambda^3/\varepsilon)(\lambda/\varepsilon)$. It is better to reorganize (3.62) as

$$\bar{Z}_\lambda(\lambda, \varepsilon) = 1 + \sum_{i=1}^\infty \frac{\lambda^i}{\varepsilon} g_i\left(\frac{\lambda}{\varepsilon}\right), \quad (3.63)$$

where the functions g_i are power series that depend only on the first i-th coefficients of the beta function. Indeed, $\lambda^i g_i/\varepsilon$ contains $i-1$ more powers of λ than powers of $1/\varepsilon$, and from (3.62) it is clear than the coefficients $\bar{\beta}_j$, $j > i$, do not fit. The *next-to-maximum* poles are those of the form $(\lambda^2/\varepsilon)(\lambda/\varepsilon)^m$, $m \geqslant 0$, the next-to-next-to-maximum poles are those of the form $(\lambda^3/\varepsilon)(\lambda/\varepsilon)^m$, $m \geqslant 0$, etc. Since the power of λ coincides with the power of $\hbar$, the poles are organized according to the general scheme

$$
\begin{array}{lllll}
L = 1 & \frac{\lambda}{\varepsilon} \\
L = 2 & \frac{\lambda^2}{\varepsilon^2} & \frac{\lambda^2}{\varepsilon} \\
L = 3 & \frac{\lambda^3}{\varepsilon^3} & \frac{\lambda^3}{\varepsilon^2} & \frac{\lambda^3}{\varepsilon} \\
L = 4 & \frac{\lambda^4}{\varepsilon^4} & \frac{\lambda^4}{\varepsilon^3} & \frac{\lambda^4}{\varepsilon^2} & \frac{\lambda^4}{\varepsilon} \\
& \cdots
\end{array}
\tag{3.64}
$$

where L is the number of loops. The i-th column is $\lambda^i g_i/\varepsilon$. The one-loop coefficient of the beta function determines the first column (i.e. the maximum poles), the one- and two-loop coefficients determine the first two columns (i.e. the maximum and the next-to-maximum poles), and so on: the j-loop coefficients, $j \leqslant i$, of the beta function determine the first i columns. At each loop, brand new information is contained only in the diagonal term. The nondiagonal elements correspond to the subdivergences.

For example, if we include the two-loop corrections to the beta function, $\beta_\lambda = \beta_1 \lambda^2 + \beta_2 \lambda^3 + \mathcal{O}(\lambda^4)$, we can determine Z_λ up to the next-to-maximum poles. We get

$$
\bar{Z}_\lambda(\lambda, \varepsilon) = \frac{1 + \frac{\beta_2 \varepsilon}{\beta_1^2} \ln\left(1 - \frac{\beta_1 \lambda}{\varepsilon}\right)}{1 - \frac{\beta_1 \lambda}{\varepsilon}} + \frac{\beta_2 \lambda}{\beta_1 \left(1 - \frac{\beta_1 \lambda}{\varepsilon}\right)^2} = \frac{1}{1 - \frac{\beta_1 \lambda}{\varepsilon}} + \frac{\lambda^2}{\varepsilon} g_2\left(\frac{\lambda}{\varepsilon}\right),
$$

up to corrections $\lambda^i g_i/\varepsilon$ with $i > 2$.

The poles of a generic correlation function G have a similar structure, where now the first i columns receive contributions from the j-loop coefficients, $j \leqslant i$, of the beta functions and the anomalous dimensions. We have, in the minimal subtraction scheme,

$$
G(\lambda, \varepsilon) = G_c + \sum_{i=1}^{\infty} \frac{\lambda^i}{\varepsilon} G_i\left(\frac{\lambda}{\varepsilon}\right),
\tag{3.65}
$$

where G_c is the tree-level contribution and G_i are power series in λ/ε. The i-th column of (3.64) is $\lambda^i G_i/\varepsilon$.

Now, assume that the first coefficient β_1 vanishes. Then (3.62) becomes

$$
\bar{Z}_\lambda(\lambda, \varepsilon) = 1 + \sum_{i=2}^{\infty} \bar{\beta}_i \frac{\lambda^i}{\varepsilon} f_i\left(\bar{\beta}_2 \frac{\lambda^2}{\varepsilon}, \cdots, \bar{\beta}_i \frac{\lambda^i}{\varepsilon}\right).
$$

The first column of the scheme (3.64) disappears, the second column is just made of its top element, and the other columns are made by their top elements and sparse other elements. Again, the first i columns are determined by the first i coefficients of the beta function. Similar restrictions apply when the first two coefficients of the beta function vanish, and so on.

Another way to reach the same conclusions is to write

$$\ln \bar{Z}_\lambda = \sum_{i=1}^{\infty} \frac{c_i(\lambda)}{\varepsilon^i},$$

where $c_i(\lambda)$ are power series in λ that begin with $\mathcal{O}(\lambda^i)$. We insert this expression into formula (3.8). Equating each order in ε we get

$$\bar{\beta}_\lambda = \sigma \lambda^2 \frac{\mathrm{d}c_1(\lambda)}{\mathrm{d}\lambda}, \qquad \frac{\mathrm{d}c_i(\lambda)}{\mathrm{d}\lambda} = \frac{\bar{\beta}_\lambda}{\sigma\lambda} \frac{\mathrm{d}c_{i-1}(\lambda)}{\mathrm{d}\lambda} \quad \text{for } i > 1.$$

Knowing that $c_i(\lambda) = \mathcal{O}(\lambda^i)$ we find

$$c_1(\lambda) = \int_0^\lambda \frac{\bar{\beta}_\lambda(\lambda')}{\sigma\lambda'^2} \mathrm{d}\lambda', \qquad c_i(\lambda) = \int_0^\lambda \frac{\mathrm{d}c_{i-1}(\lambda')}{\mathrm{d}\lambda'} \frac{\bar{\beta}_\lambda(\lambda')}{\sigma\lambda'} \mathrm{d}\lambda' \quad \text{for } i > 1.$$

We see that the one-loop beta function determines the orders $\mathcal{O}(\lambda^i)$ of all the $c_i(\lambda)$'s. Similarly, the two-loop beta function determines all the $c_i(\lambda)$'s up to and including $\mathcal{O}(\lambda^{i+1})$, and the k-loop beta function determines them up to and including $\mathcal{O}(\lambda^{i+k-1})$.

To summarize, the power of the renormalization group is that it relates infinitely many quantities, such as the entries of the columns of (3.64), and allows us to resum them. A consequence is that computing the various entries of a column involves more or less the same level of difficulty. As a check of this, we suggest the reader to compute the two-loop double poles of Z_λ in the massless φ_4^4 theory, which is part of exercise 6. It will be easily realized that the calculation requires an effort that is considerably smaller than the one required by the full exercise, and is comparable to the typical effort of one-loop calculations. Similar simplifications occur with the three-loop triple poles, and so on.

To further appreciate the meaning of these facts, consider the formula (3.53) of the one-loop running coupling, and compare it to the formula of the bare coupling:

$$\tilde{\lambda}(t, \lambda) = \frac{\lambda}{1 - \frac{3t\lambda}{16\pi^2}}, \qquad \lambda_B \mu^{-\varepsilon} = \frac{\lambda}{1 - \frac{3\lambda}{16\pi^2\varepsilon}}. \qquad (3.66)$$

We see that $\lambda_B \mu^{-\varepsilon}$ is nothing but the running coupling $\tilde{\lambda}(t, \lambda)$ at $t = 1/\varepsilon$. In a cutoff regularization framework, it would be the running coupling at an energy scale equal to the cutoff Λ, since $1/\varepsilon \sim \ln(\Lambda/\mu)$. The resummation of the powers of λ/ε in $\lambda_B \mu^{-\varepsilon}$ is just like the resummation of the powers of $t\lambda$ in $\tilde{\lambda}(t, \lambda)$.

We can better understand the properties of the RG resummations by studying how they work inside the correlation functions. Assuming, for definiteness, that the theory contains a unique field φ and a unique dimensionless coupling λ, let us consider

$$W(x, \lambda, \mu) = \langle \varphi(x_1) \cdots \varphi(x_n) \rangle_c = \exp\left(n \int_\lambda^{\tilde{\lambda}(t)} d\lambda' \frac{\gamma(\lambda')}{\beta_\lambda(\lambda')} \right) W(x, \tilde{\lambda}(t), \tilde{\mu}).$$

(3.67)

We have used the solution (3.36) of the Callan-Symanzik equation, set $\varepsilon = 0$, and rewritten the left formula of (3.37) by means of the integral on the coupling. The limit of integration λ that appears on the right-hand side of (3.67) has no physical meaning, since it can be absorbed into the normalization of the field. The cross sections and the other physical quantities do not depend on such a normalization. Ignoring it, the right-hand side of (3.67) depends on λ just through the running coupling $\tilde{\lambda}(t)$. If $\tilde{\lambda}$ is small, the perturbative expansion of the right-hand side in powers of $\tilde{\lambda}$ makes sense, and can be physically useful.

Formula (3.46) shows that when $|t|$ is large, the running coupling can be small, even if λ is of order one. The point is that if λ is of order one, the perturbative expansion of the left-hand side of (3.67) is purely formal: it does not converge, and is practically useless! So, there are situations where the right-hand side makes sense, while the left-hand side does not make sense (and vice versa). Given that the two expressions are equal, what is happening?

What happens is that the renormalization group teaches us how to *reorganize* the expansion appearing to the left as the expansion appearing to the right. What we have to do is to *formally* resum the powers of λ into $\tilde{\lambda}$, ignoring the fact that such a sum may not be convergent.

We have already remarked that little is known about the resummation of the perturbative expansion in quantum field theory, to the extent that different ways to organize the sums may lead to different results, and physically different theories. The renormalization group comes to the rescue, to some extent. We start with a generic purpose of defining a formal perturbative expansion in powers of λ, because we hope that it may be physically useful when λ is small. The first few orders may be sufficiently accurate for a prediction, if we have reasons to discard the higher-order corrections. That is all we are asking in the first place.

Yet, we discover along the way that we can make sense of expansions that appear to be meaningless at first. Or we can turn an expansion that is useful in certain situations, into an expansion that is useful in different situations. Gathering the powers of $t\lambda$ appropriately, as in the left expression of (3.66), we turn an expansion in powers of λ, which is useful when λ is small, into an expansion in powers of $\tilde{\lambda}$, which is useful when $\tilde{\lambda}$ is small. There might be no relation between the two situations, to the extent that the condition $\tilde{\lambda} \ll 1$ typically singles out the ultraviolet limit or the infrared one. This is possible because of the nontrivial "conceptual jump" involved in the switch from the left to the right of equation (3.67): the resummation of a λ expansion that may not be convergent.

Ultimately, renormalization is so selective that it guides us towards what makes sense and what must be discarded, what we can do and what we cannot do: it is not in our hands, any longer. Who could foresee that the strong interactions were meant to become weak at high energies? It is just like that, and there is no parameter we can tune to make it not happen.

Chapter 4

Gauge symmetry

In this chapter we study Abelian and non-Abelian gauge symmetries in quantum field theory. After giving the basic notions and the main properties, we discuss the problems raised by their quantization, such as the gauge fixing and unitarity. In the next chapter we upgrade the formalism to make it suitable to prove the renormalizability of gauge theories to all orders. Then we proceed by proving the renormalizability of quantum electrodynamics (chapter 6) and the renormalizability of non-Abelian gauge field theories (chapter 7). We assume that the theory is parity invariant, so no chiral fermions are present. The renormalization of parity violating quantum field theories raises other issues.

4.1   Abelian gauge symmetry

The propagation of free massless vector fields A_μ is described by the massless limit of the Proca action (1.89),

$$S_{\text{free}}(A) = \frac{1}{4} \int \mathrm{d}^D x \, F_{\mu\nu}^2, \tag{4.1}$$

where $F_{\mu\nu} = \partial_\mu A_\nu - \partial_\nu A_\mu$ is the field strength. This action is invariant under the gauge symmetry

$$A'_\mu = A_\mu + \partial_\mu \Lambda, \tag{4.2}$$

where Λ is an arbitrary function of the spacetime point. In infinitesimal form, the symmetry transformation reads

$$\delta_\Lambda A_\mu = \partial_\mu \Lambda. \tag{4.3}$$

We have already written the action (4.1) in complex D dimensions, because one of the main virtues of the dimensional-regularization technique is that it is manifestly gauge invariant, as long as the theory does not contain chiral fermions. Gauge invariance looks formally the same in all (integer) dimensions, so it is easy to generalize it to the formal objects A_μ, ∂_μ, x^μ, γ^μ, ψ, etc., which are used in the context of the dimensional regularization. Instead, the notion of chirality is dimension dependent, so gauge invariance is not manifest in D dimensions when the Lagrangian explicitly contains γ_5.

A direct consequence of the local gauge symmetry is that the quadratic part of the action (4.1) is not invertible. Indeed, it is proportional to $k^2 \delta_{\mu\nu} - k_\mu k_\nu$ in momentum space, which has the null eigenvector k_ν. Therefore, the Green function $\langle A_\mu(k) A_\nu(-k) \rangle$ is not well defined. This fact is also evident from the massless limit of the Proca propagator (1.91), which is singular.

The action (1.104) of free fermions is invariant under the global $U(1)$ transformation

$$\psi' = \mathrm{e}^{-ie\Lambda} \psi, \tag{4.4}$$

where Λ is constant. If we extend Λ to a function $\Lambda(x)$ of the spacetime point, (1.104) is no longer invariant, because of a correction generated by the derivative of ψ. Nevertheless, if we pair the partial derivative ∂_μ to a field that transforms so as to compensate the correction in question, we can extend the symmetry as desired. That field is a vector transforming as (4.2).

The photon is the gauge field A_μ that promotes the global $U(1)$ invariance (4.4) to a local symmetry

$$A'_\mu = A_\mu + \partial_\mu \Lambda, \qquad \psi' = \mathrm{e}^{-ie\Lambda} \psi, \qquad \bar\psi' = \mathrm{e}^{ie\Lambda} \bar\psi. \tag{4.5}$$

Replacing the simple derivative ∂_μ with the covariant derivative $\partial_\mu + ieA_\mu$ and adding (4.1), we obtain the Lagrangian of quantum electrodynamics (QED)

$$\mathcal{L}_0 = \frac{1}{4} F_{\mu\nu}^2 + \bar\psi (\slashed{\partial} + ie\slashed{A} + m)\psi, \tag{4.6}$$

where $\rlap{/}{A} = A_\mu \gamma^\mu$, which is invariant under the gauge transformation (4.5). In infinitesimal form, (4.5) becomes

$$\delta_\Lambda A_\mu = \partial_\mu \Lambda, \qquad \delta_\Lambda \psi = -ie\Lambda\psi, \qquad \delta_\Lambda \bar\psi = ie\Lambda\bar\psi. \tag{4.7}$$

4.2   Gauge fixing

To define the functional integral of a gauge theory, and its perturbative expansion, we first need to choose a gauge, by imposing a condition of the form

$$\mathcal{G}(A) = 0,$$

where $\mathcal{G}(A)$ is a suitable local function. We will have to show that the physical quantities do not depend on the choice we make. Among the most popular gauge choices we mention the Lorenz gauge fixing

$$\mathcal{G}(A) = \partial_\mu A_\mu, \tag{4.8}$$

and the Coulomb gauge fixing

$$\mathcal{G}_C(A) = -\boldsymbol{\nabla} \cdot \mathbf{A}. \tag{4.9}$$

To implement the gauge fixing at the quantum level, we start from the generating functional

$$Z(J) = \int [\mathrm{d}A] \exp\left(-S(A) + \int JA\right) \tag{4.10}$$

in the absence of matter, where $S(A)$ is the free action (4.1) and JA stands for $J_\mu A_\mu$. The functional measure is gauge invariant, since the gauge transformation is just a translation. If the current J is divergenceless, $Z(J)$ is formally gauge invariant. Then, however, we know that it is ill defined, because the integrand is independent of the longitudinal mode (4.3), on which we are integrating. We may have a chance of defining $Z(J)$ if we keep the current J generic. The problem is that a generic J does not seem compatible with gauge invariance. We are going to discover in a minute that it actually is.

Let us insert "1", written in the form

$$1 = \int [\mathrm{d}\Lambda] \, (\det \Box) \, \delta_F(-\partial_\mu A_\mu + Q + \Box\Lambda), \tag{4.11}$$

where Q is an arbitrary function. Here $\delta_F(Y)$ denotes the "functional delta function", which means the product of $\delta(Y(x))$ over all the spacetime points x, where $Y(x)$ is a generic function (or an operator, as in the present case). Formula (4.11) is the functional version of the ordinary formula

$$\int \prod_{i=1}^{n} \mathrm{d}x_i \, \mathcal{J}(x) \prod_{k=1}^{n} \delta(f_k(x)) = \sum_{\bar{x}} \mathrm{sgn}(\mathcal{J}(\bar{x})). \qquad (4.12)$$

where

$$\mathcal{J}(x) = \det \frac{\partial f_i(x)}{\partial x_j},$$

$\bar{x}$ are the points where the functions $f_k(x)$ vanish simultaneously, and sgn is the sign function $\mathrm{sgn}(x) = x/|x|$. We assume that the $\bar{x}$ are distinct.

As long as it does not vanish, the right-hand side of (4.12) is just a normalization factor, which can be omitted. In our case it is precisely 1, so we get

$$Z(J) = \int [\mathrm{d}A\mathrm{d}\Lambda] \, (\det \Box) \delta_F(-\partial_\mu A_\mu + Q + \Box\Lambda) \exp\left(-S(A) + \int JA\right).$$

Now, perform a change of variables $A' = A - \partial\Lambda$. Recalling that the functional measure is invariant under translations, we obtain

$$Z(J) = \int [\mathrm{d}A\mathrm{d}\Lambda] \, (\det \Box) \delta_F(-\partial_\mu A_\mu + Q) \exp\left(-S(A) + \int (JA - \Lambda\partial \cdot J)\right),$$
$$(4.13)$$

after dropping the primes. We see that the integral over the longitudinal mode Λ factorizes the expression

$$\int [\mathrm{d}\Lambda] \exp\left(-\int \mathrm{d}^D x \Lambda(x)(\partial \cdot J(x))\right).$$

It is easier to interpret this expression once we replace Λ with $i\Lambda$. Operations like this are allowed, because they have no impact on the diagrammatics. We must keep in mind that we are working perturbatively, so the functional integral must be understood as a book-keeping for the collection of diagrams.

Omitting overall constant factors, which are also irrelevant, we get a functional delta function that enforces the current conservation:

$$\prod_x \int \mathrm{d}\Lambda_x e^{-i\Lambda_x(\partial_\mu J_\mu(x))} = \prod_x \delta_F(\partial_\mu J_\mu(x)) = \delta_F(\partial \cdot J).$$

In other words, we obtain that J is divergenceless. We obtain it for free, without having to assume it from the start! This means that, despite its appearance, (4.10) is gauge invariant for arbitrary J.

Going back to (4.13), we find

$$Z(J) = \delta_F(\partial \cdot J) \int [\mathrm{d}A] \, (\det \Box) \delta_F(-\partial_\mu A_\mu + Q) \exp\left(-S(A) + \int JA\right),$$

which is now gauge-fixed. At this point, we introduce a "Lagrange multiplier" B[1] and write the functional delta function appearing in the integrand as

$$\delta_F(-\partial_\mu A_\mu + Q) = \int [\mathrm{d}B] \exp\left(-i \int B(\partial_\mu A_\mu - Q)\right).$$

We prefer to work with Hermitian quantities in the exponent, so we replace B with $-iB$. As explained above, the diagrammatic meaning of the functional integral ensures that this operation is permitted. Then,

$$Z(J) = \delta_F(\partial \cdot J) \int [\mathrm{d}\bar{\mu}] \, (\det \Box) \exp\left(-\int \left(\frac{1}{4}F_{\mu\nu}^2 + B\partial_\mu A_\mu - BQ - JA\right)\right),$$
$$(4.14)$$

where $[\mathrm{d}\bar{\mu}] = [\mathrm{d}A\mathrm{d}B]$. We see that the function Q plays the role of an external source for the Lagrange multiplier B. We can easily work out the propagator of the pair (A_μ, B). The result is

$$\begin{pmatrix} \langle A_\mu(k)A_\nu(-k)\rangle & \langle A_\mu(k)B(-k)\rangle \\ \langle B(k)A_\nu(-k)\rangle & \langle B(k)B(-k)\rangle \end{pmatrix} = \frac{1}{k^2} \begin{pmatrix} \delta_{\mu\nu} - \frac{k_\mu k_\nu}{k^2} & -ik_\mu \\ ik_\nu & 0 \end{pmatrix}. \quad (4.15)$$

The functional integral can be evaluated straightforwardly, and gives

$$Z(J) = \delta_F(\partial \cdot J)(\det \Box)^{-1} \exp\left(\frac{1}{2} \int \mathrm{d}^D x \, J_\mu(x) G_A(x-y) J_\mu(y) \mathrm{d}^D y\right),$$

where the Green function $G_A(x-y)$ coincides with the bosonic one of formula (1.36) at $m = 0$, or the expression (1.37) with $x \to x - y$. In the exponent, we have used that J is divergenceless. The result is independent of Q, which was expected, since Q is arbitrary in formula (4.11).

Formula (4.14) contains a $\det \Box$ in the numerator, for which it is not easy to write Feynman rules. We could ignore this factor, because it is just

---

[1] Also known as Nakanishi-Lautrup auxiliary field.

a constant in QED. However, in more general gauge theories the analogue of this factor depends on the fields. If we introduce extra, anticommuting fields C and $\bar{C}$, called Faddeev-Popov *ghosts* and *antighosts*, respectively, we can exponentiate the determinant by means of formula (1.95). The complete gauge-fixed generating functional then reads

$$Z(J) = \delta_F(\partial \cdot J) \int [\mathrm{d}\tilde{\mu}] \exp\left(-\int \left(\frac{1}{4}F_{\mu\nu}^2 + B\partial_\mu A_\mu - \bar{C}\Box C - BQ - JA\right)\right),$$
(4.16)

where $[\mathrm{d}\tilde{\mu}] = [\mathrm{d}A\mathrm{d}C\mathrm{d}\bar{C}\mathrm{d}B]$. The ghost propagator is simply

$$\langle C(k)\bar{C}(-k)\rangle = \frac{1}{k^2}.$$
(4.17)

There is no need to keep the overall factor $\delta_F(\partial \cdot J)$ enforcing the current conservation, since we can always restore this condition at a later time. Thus, from now on we drop that factor, and relax the assumption that J is divergenceless.

If we average over Q with the Gaussian measure

$$\int [\mathrm{d}Q] \exp\left(-\frac{1}{2\lambda}\int \mathrm{d}^D x\, Q(x)^2\right),$$

where λ is an arbitrary parameter, we obtain

$$Z'(J) = \int [\mathrm{d}\tilde{\mu}] \exp\left(-\int \left(\frac{1}{4}F_{\mu\nu}^2 - \frac{\lambda}{2}B^2 + B(\partial \cdot A) - \bar{C}\Box C - JA\right)\right).$$
(4.18)

Since B now appears quadratically, it can be easily integrated away, giving

$$Z'(J) = \int [\mathrm{d}\mu] \exp\left(-\int \left(\frac{1}{4}F_{\mu\nu}^2 + \frac{1}{2\lambda}(\partial \cdot A)^2 - \bar{C}\Box C - JA\right)\right),$$
(4.19)

where $[\mathrm{d}\mu] = [\mathrm{d}A\mathrm{d}C\mathrm{d}\bar{C}]$.

In this framework, the ghost propagator (4.17) is unchanged, while the gauge-field propagator reads

$$\langle A_\mu(k)\, A_\nu(-k)\rangle = \frac{1}{k^2}\left(\delta_{\mu\nu} + (\lambda - 1)\frac{k_\mu k_\nu}{k^2}\right).$$
(4.20)

A simple gauge choice is the *Feynman gauge*, where $\lambda = 1$ and

$$\langle A_\mu(k)\, A_\nu(-k)\rangle = \frac{\delta_{\mu\nu}}{k^2}.$$
(4.21)

The choice $\lambda = 0$ is also known as the *Landau gauge*.

Everything we said so far can be repeated by replacing $\partial_\mu A_\mu$ in (4.11) with the Coulomb gauge fixing (4.9). Then we get

$$Z'(J) = \int [d\tilde{\mu}] \, \exp\left(-\int \left(\frac{1}{4}F_{\mu\nu}^2 - \frac{\lambda}{2}B^2 - B\nabla \cdot \mathbf{A} + \bar{C}\triangle C - JA\right)\right),$$
(4.22)

and, after integrating B away,

$$Z'(J) = \int [d\mu] \, \exp\left(-\int \left(\frac{1}{4}F_{\mu\nu}^2 + \frac{1}{2\lambda}(\nabla \cdot \mathbf{A})^2 + \bar{C}\triangle C - JA\right)\right). \quad (4.23)$$

For the moment we content ourselves with these two choices of gauge fixing. However, the gauge-fixing function $\mathcal{G}(A)$ can in principle be arbitrary, as long as it properly fixes the gauge.

Physical degrees of freedom

The physical degrees of freedom are more clearly visible in the Coulomb gauge. The gauge-field propagators of (4.23) read, in Minkowski spacetime,

$$\langle A_0(k)A_0(-k)\rangle_{\mathrm{M}} = \frac{1}{\mathbf{k}^2} - \frac{\lambda E^2}{(\mathbf{k}^2)^2}, \qquad \langle A_0(k)A_i(-k)\rangle_{\mathrm{M}} = -\frac{\lambda E k_i}{(\mathbf{k}^2)^2},$$

$$\langle A_i(k)A_j(-k)\rangle_{\mathrm{M}} = \frac{1}{E^2 - \mathbf{k}^2}\left(\delta_{ij} - \frac{k_i k_j}{\mathbf{k}^2}\right) - \frac{\lambda k_i k_j}{(\mathbf{k}^2)^2}. \quad (4.24)$$

To switch from the Euclidean framework to the Minkowskian one, we have written $A = (iA_0, \mathbf{A})$ and $k = (iE, \mathbf{k})$, and then recalled that the Fourier transform of a field gets a further factor i. Studying the poles of (4.24), we see that only $\langle A_i A_j \rangle$ has any, precisely two. They have positive residues, and their dispersion relations are $E = |\mathbf{k}|$. The ghost propagator is now

$$\langle C(k)\bar{C}(-k)\rangle_{\mathrm{M}} = \frac{1}{\mathbf{k}^2} \quad (4.25)$$

and has no pole. In total, the physical degrees of freedom are 2, as expected.

In the Lorenz gauge the propagators have more complicated pole structures. For example, the ghost propagator (4.17) has poles and the gauge-field propagators (4.20) and (4.21) have extra poles. We will show that the unphysical degrees of freedom that appear with an arbitrary choice of gauge fixing compensate one another. More precisely, the physical quantities do not depend on the gauge fixing, which allows us to switch to the Coulomb gauge, where no unphysical poles are present.

When we add matter, the gauge-fixing procedure does not change. For example, in the Lorentz gauge the gauge-fixed Lagrangian of QED is

$$\mathcal{L}_{\text{gf}} = \frac{1}{4}F_{\mu\nu}^2 + \bar{\psi}(\slashed{\partial} + ie\slashed{A} + m)\psi - \frac{\lambda}{2}B^2 + B\,\partial\cdot A - \bar{C}\Box C,$$

before integrating B out. For completeness, we report the propagator of the multiplet made of A_μ and B

$$\begin{pmatrix} \langle A_\mu(k)A_\nu(-k)\rangle & \langle A_\mu(k)B(-k)\rangle \\ \langle B(k)A_\nu(-k)\rangle & \langle B(k)B(-k)\rangle \end{pmatrix} = \frac{1}{k^2}\begin{pmatrix} \delta_{\mu\nu} + (\lambda-1)\dfrac{k_\mu k_\nu}{k^2} & -ik_\mu \\ ik_\nu & 0 \end{pmatrix},$$
$$(4.26)$$

which coincides with (4.15) at $\lambda = 0$.

The gauge-fixing procedure we have described breaks the local symmetry (4.3). Nevertheless, the symmetry is not truly lost, because the functional integral acquires new properties. Thanks to those, we will be able to prove that the physical quantities are gauge invariant and gauge independent, before and after renormalization. Such properties are elegantly combined in a very practical and compact canonical formalism. That formalism is actually more than we need for Abelian theories, but has the virtue of providing a unified treatment that is also suitable to treat the non-Abelian gauge theories, as well as quantum gravity and every general gauge theory. Before laying out the canonical formalism, we introduce the non-Abelian gauge symmetry.

4.3   Non-Abelian global symmetry

Consider a multiplet ψ^i of fermionic fields. The free Lagrangian

$$\sum_i \left(\bar{\psi}^i \slashed{\partial}\psi^i + m\bar{\psi}^i\psi^i \right)$$

is invariant under the global transformation

$$\psi^{i\prime} = U^{ij}\psi^j, \qquad \bar{\psi}^{i\prime} = \bar{\psi}^j U^{\dagger ji}, \tag{4.27}$$

where U is a unitary matrix. More generally, given a non-Abelian unitary group G, we can consider multiplets ψ^i of fermionic fields that transform according to some representation of G, and theories that are symmetric with respect to these global transformations.

We focus our attention on the case $G = SU(N)$, where U can be parametrized as

$$U = \exp\left(-g\Lambda^a T^a\right), \tag{4.28}$$

by means of a basis of $N \times N$ traceless anti-Hermitian matrices T_{ij}^a, where g and Λ^a are real constants, and a is an index ranging from 1 to $\dim G = N^2 - 1$.

Consider the commutator $[T^a, T^b]$: since it is traceless and anti-Hermitian, it can be expanded in the basis T^a. We have

$$[T^a, T^b] = f^{abc} T^c, \tag{4.29}$$

where f^{abc} are real numbers such that

$$f^{abc} = -f^{bac}, \tag{4.30}$$

$$f^{abc} f^{cde} + f^{dac} f^{cbe} + f^{bdc} f^{cae} = 0. \tag{4.31}$$

The second line follows from the Jacobi identity obeyed by the commutator. The matrices T^a can be normalized so that

$$\mathrm{tr}[T^a T^b] = -\frac{1}{2}\delta^{ab}, \tag{4.32}$$

where the sign is determined by the anti-Hermiticity, while the factor $1/2$ is conventional. In a basis where (4.32) holds, the constants f^{abc} are completely antisymmetric, which can be proved from

$$\mathrm{tr}[T^c[T^a, T^b]] = -\frac{1}{2}f^{abc}, $$

by using the cyclicity of the trace.

For example, in the case $G = SU(2)$ we have $T^a = -i\sigma^a/2$, where σ^a are the Pauli matrices in the standard basis, and $f^{abc} = \varepsilon^{abc}$.

Any set of real constants f^{abc} that satisfy the properties (4.30) and (4.31) defines a *Lie algebra*. The f^{abc}'s are called *structure constants* of the algebra. We can introduce abstract *generators* $\mathcal{T}^a$ that satisfy the formal commutation rules

$$[\mathcal{T}^a, \mathcal{T}^b] = f^{abc} \mathcal{T}^c. \tag{4.33}$$

When the generators T^a are given an explicit form, as matrices acting on the vectors of some vector space V, we have a *representation* of the Lie algebra. Here we consider only finite dimensional representations. A representation is irreducible if the whole of V is filled by repeatedly acting with the T^a's on any nontrivial vector of V.

The set of $N \times N$ traceless anti-Hermitian matrices T^a form the *fundamental* representation of the Lie algebra of $SU(N)$, which is the one of minimal dimension greater than one. It is commonly denoted by its very dimension, N. Taking the complex conjugate of (4.29), we obtain another representation with $T^a = \bar{T}^a$, called *antifundamental*, commonly denoted by $\bar{N}$. The trivial representation, which has dimension 1, is called *singlet*.

In a generic irreducible representation r, the matrices T^a can be normalized so that

$$\mathrm{tr}[T^a T^b] = -C(r)\delta^{ab}, \tag{4.34}$$

where $C(r)$ is a constant depending on r. We have $C(N) = C(\bar{N}) = 1/2$. Another important result follows from the observation that $T^a T^a$ commutes with every generator T^a, hence it is a constant on every irreducible representation. We write

$$T^a T^a = -C_2(r)\mathbb{1}, \tag{4.35}$$

where $C_2(r)$ is called the *quadratic Casimir element* of the representation r.

A consequence of the Jacobi identity (4.31) is that the matrices

$$(\tau^a)^{bc} = -f^{abc}$$

satisfy the commutation rules (4.33), so they form another representation of the Lie algebra, called *adjoint representation*, normally denoted by G. Contracting a and b in (4.34) and tracing the equation (4.35), we get

$$C(r)d(G) = C_2(r)d(r),$$

where $d(r)$ is the dimension of the representation r. With $r = N$ we find $C_2(N) = (N^2 - 1)/(2N)$. Choosing $r = G$ we obtain that the two constants of the adjoint representation coincide: $C(G) = C_2(G)$. It can be shown that $C(G) = C_2(G) = N$. In particular, using (4.34) for $r = G$, we get

$$f^{acd} f^{bcd} = N\delta^{ab}. \tag{4.36}$$

Observe that any $N \times N$ matrix can be written as a complex linear combination of the identity matrix and the matrices T^a. Consider the tensor $\delta_{ij}\delta_{kl}$, and view it as a collection of $N \times N$ matrices with indices j and k. It can be expanded as

$$\delta_{ij}\delta_{kl} = \alpha_{il}\delta_{kj} + \alpha_{il}^a T_{jk}^a, \tag{4.37}$$

where α_{il} and α_{il}^a are complex numbers. Taking $j = k$ we get

$$\delta_{il} = N\alpha_{il}. \tag{4.38}$$

Moreover, we also have

$$T_{li}^a = T_{kj}^a \delta_{ij}\delta_{kl} = \alpha_{il}^b \operatorname{tr}[T^a T^b] = -\frac{1}{2}\alpha_{il}^a. \tag{4.39}$$

Collecting (4.38) and (4.39), formula (4.37) can be rephrased as

$$T_{ij}^a T_{kl}^a = -\frac{1}{2}\left(\delta_{il}\delta_{kj} - \frac{1}{N}\delta_{ij}\delta_{kl}\right). \tag{4.40}$$

This identity is often useful.

We have started this discussion from fields ψ^i and $\bar{\psi}^i$ which, by (4.27), transform according to the fundamental and the antifundamental representations, respectively. It is convenient to introduce a notation that distinguishes the indices of these two types. Let

$$v_{j_1 \cdots j_m}^{i_1 \cdots i_n}$$

denote a tensor whose upper n indices transform according to the fundamental representation and whose lower m indices transform according to the antifundamental representation. Globally and infinitesimally, we have

$$v_{j_1 \cdots j_m}^{\prime i_1 \cdots i_n} = U_{j_1}^{\dagger l_1} \cdots U_{j_m}^{\dagger l_m} U_{k_1}^{i_1} \cdots U_{k_n}^{i_n} v_{l_1 \cdots l_m}^{k_1 \cdots k_n},$$

and

$$\delta_\Lambda v_{j_1 \cdots j_m}^{i_1 \cdots i_n} = -g \sum_{a=1}^{\dim G} \Lambda^a \left(T^{a i_1}{}_{l_1} v_{j_1 \cdots j_m}^{l_1 i_2 \cdots i_n} + \cdots + T^{a i_n}{}_{l_n} v_{j_1 \cdots j_m}^{i_1 \cdots i_{n-1} l_n} \right)$$

$$+ g \sum_{a=1}^{\dim G} \Lambda^a \left(T^{a k_1}{}_{j_1} v_{k_1 j_2 \cdots j_m}^{i_1 \cdots i_n} + \cdots + T^{a k_m}{}_{j_m} v_{j_1 \cdots j_{m-1} k_m}^{i_1 \cdots i_n} \right), \tag{4.41}$$

respectively. We have written the matrices T_{ij}^a as $T^{ai}{}_j$ to emphasize the roles of their indices.

The tensors

$$\delta_j^i, \qquad \varepsilon^{i_1\cdots i_N}, \qquad \varepsilon_{i_1\cdots i_N}, \tag{4.42}$$

are clearly invariant. Observe that δ_j^i can contract only different types of indices.

Let u^i, v^i, ... and u_i, v_i,... denote vectors transforming in the fundamental and antifundamental representations, respectively. We can construct new representations by considering the products $u^i v_j w^k \cdots$. Using the tensors (4.42), the products of fundamental and antifundamental representations can be decomposed into sums of irreducible representations. The decomposition is obtained by repeatedly subtracting contributions proportional to the invariant tensors, until what remains vanishes whenever it is contracted with invariant tensors.

For example, the product $u^i v_j$ of a fundamental and an antifundamental representation can be decomposed as the sum of two irreducible representations, as follows

$$u^i v_j = \left(P_{1jk}^{il} + P_{2jk}^{il} \right) u^k v_l, \tag{4.43}$$

by means of the projectors

$$P_{1jk}^{il} = \delta_k^i \delta_j^l - \frac{1}{N} \delta_j^i \delta_k^l, \qquad P_{2jk}^{il} = \frac{1}{N} \delta_j^i \delta_k^l. \tag{4.44}$$

It can be shown that the traceless combination given by P_1 is equivalent to the adjoint representation. The term proportional to δ_j^i is obviously a singlet. We symbolically write the decomposition (4.43) as

$$N \times \bar{N} = \mathrm{Adg} + \mathbb{1}.$$

The matrices $\mathcal{T}^{ail}{}_{jk}$ of the representation acting on the product $u^k v_l$ are $T^{ai}{}_k \delta_j^l - T^{al}{}_j \delta_k^i$. They do not need to be projected with (4.44), since they act nontrivially only on the adjoint combination.

Another example is the product $u^i v^j$ of two fundamental representations. It can be decomposed as the sum of the symmetric and antisymmetric combinations,

$$u^i v^j = \frac{1}{2}(u^i v^j + u^i v^j) + \frac{1}{2}(u^i v^j - u^i v^j),$$

which are new representations of dimensions $N(N+1)/2$ and $N(N-1)/2$, respectively. We have

$$T_{\pm}^{aij}{}_{kl} = \frac{1}{2}\left(T^{ai}{}_k \delta^j_l + \delta^i_k T^{aj}{}_l \pm T^{aj}{}_k \delta^i_l \pm \delta^j_k T^{ai}{}_l\right),$$

or, symbolically,

$$T_{\pm}^a = P_{\pm}\left(T^a \otimes \mathbb{1} + \mathbb{1} \otimes T^a\right)P_{\pm},$$

where $P_{\pm}$ are the projectors on the symmetric and antisymmetric combinations, respectively.

A theorem ensures that all the representations can be obtained by means of a similar procedure.

Theorem 4 *All the irreducible finite dimensional representations can be obtained from the products of fundamental and antifundamental representations, decomposed by means of the invariant tensors (4.42).*

Actually, even the antifundamental representation can be obtained from the fundamental one. Indeed,

$$u_i \equiv \frac{1}{(N-1)!}\varepsilon_{ik_2\cdots k_N} v^{k_2}\cdots v^{k_N}$$

does transform in the antifundamental representation. Thus,

Theorem 5 *All the irreducible finite dimensional representations can be obtained by decomposing products of fundamental representations.*

The theorem just stated ensures that

Corollary 6 *the generators T^a of every representation can be written using the matrices $T^{ai}{}_j$ and the invariant tensors (4.42).*

Symbolically, we can write

$$T_r^a = P_r(T^a \otimes \mathbb{1} \cdots \otimes \mathbb{1} + \cdots + \mathbb{1} \otimes \cdots \mathbb{1} \otimes T^a)P_r, \qquad (4.45)$$

where P_r is the projector on the representation r, constructed with the tensors (4.42).

Observe that the matrices $T^{ai}{}_j$ can be viewed as invariant tensors. Indeed, they do not change when we transform them as dictated by their three indices (fundamental representation for i, antifundamental representation for j, adjoint representation for a). They can be used as tools to convert indices into indices of different types. Formula (4.45) implies that the matrices T_r^a are also invariant tensors.

Expanding by means of the Kronecker tensor and using the identities already proved, we also find

$$f^{abc}T^{ai}{}_j T^{bk}{}_l T^{cm}{}_n = \frac{1}{4}(\delta^i_l \delta^k_n \delta^m_j - \delta^i_n \delta^k_j \delta^m_l). \tag{4.46}$$

4.4   Non-Abelian gauge symmetry

Now we gauge the non-Abelian symmetry. We promote the matrix (4.28) to a family of spacetime-dependent unitary matrices

$$U(x) = e^{-g\Lambda^a(x)T^a}. \tag{4.47}$$

Furthermore, we introduce non-Abelian gauge fields A_μ, with which we build the covariant derivative

$$(D_\mu \psi)^i = \partial_\mu \psi^i + ig(A_\mu \psi)^i, \tag{4.48}$$

where g denotes the gauge coupling. Formula (4.48) shows that the A_μ's must be matrices with indices ij. The gauge field A_μ is often called gauge connection, or gauge potential.

We determine the transformation A'_μ of A_μ by requiring that $(D_\mu \psi)^i$ transforms exactly as ψ^i. We have

$$(D_\mu \psi)'^i = \partial_\mu \psi'^i + ig(A'_\mu \psi')^i$$
$$= (\partial_\mu U^{ij})\psi^j + ig(A'_\mu U - U A_\mu)^{ij}\psi^j + U^{ij}D_\mu \psi^j = U^{ij}D_\mu \psi^j,$$

hence

$$A'_\mu = \frac{i}{g}(\partial_\mu U)U^{-1} + U A_\mu U^{-1}, \tag{4.49}$$

The transformation rule for D_μ is then

$$D'_\mu = U D_\mu U^{-1}. \tag{4.50}$$

Since the covariant differential operator iD_μ must be Hermitian, as is $i\partial_\mu$, the matrices A_μ are also Hermitian, so they can be parametrized as

$$A^{ij}_\mu(x) = -iA^a_\mu(x)T^a_{ij}, \tag{4.51}$$

where $A_\mu^a(x)$ are real functions. We can write $D_\mu = \partial_\mu + igA_\mu$ and $D_\mu^{ij} = \delta^{ij}\partial_\mu + igA_\mu^{ij}$, in matrix and index notations, respectively. Distinguishing upper and lower indices, we have $A_{\mu\,j}^i = -iT^{ai}{}_j A_\mu^a$, which emphasizes that A_μ is the traceless product between a fundamental and an antifundamental representation. That is to say, it belongs to the adjoint representation.

Define the field strength

$$F_{\mu\nu} = \frac{1}{g}[D_\mu, D_\nu] \equiv F_{\mu\nu}^a T^a.$$

Clearly, (4.50) implies the transformation rule

$$F_{\mu\nu}' = UF_{\mu\nu}U^{-1}. \tag{4.52}$$

We find the components

$$F_{\mu\nu}^a = \partial_\mu A_\nu^a - \partial_\nu A_\mu^a + gf^{abc}A_\mu^b A_\nu^c.$$

So far, we have taken matter fields ψ^i in the fundamental representation. For fields χ^I in a generic representation r, described by matrices T_r^a, we have $A_\mu^{IJ} = -iA_\mu^a T_{r\,IJ}^a$ and still $D_\mu = \partial_\mu + igA_\mu$, while $D_\mu^{IJ} = \delta^{IJ}\partial_\mu + igA_\mu^{IJ}$.

The infinitesimal forms of the transformations (4.27), (4.52) and (4.49) read

$$\delta_\Lambda A_\mu^a = \partial_\mu\Lambda^a + gf^{abc}A_\mu^b\Lambda^c \equiv D_\mu\Lambda^a,$$
$$\delta_\Lambda F_{\mu\nu}^a = gf^{abc}F_{\mu\nu}^b\Lambda^c, \qquad \delta_\Lambda\chi^I = -gT_{r\,IJ}^a\Lambda^a\chi^J. \tag{4.53}$$

If we view the functions Λ^a as elements of the adjoint representation of G, $\delta_\Lambda A_\mu^a$ is just the covariant derivative of Λ^a.

The gauge-invariant action of the fermionic fields χ^I can be constructed by means of the covariant derivative, and reads

$$S_\psi = \int d^D x\,\bar\chi(\slashed{D} + m)\chi = \int d^D x\,\bar\chi^I(\delta_{IJ}\slashed{\partial} + g\slashed{A}^a T_{r\,IJ}^a + m\delta_{IJ})\chi^J. \tag{4.54}$$

Similarly, if φ are (complex) scalar fields transforming according to some representation of the gauge group, the scalar action

$$S_\varphi = \int d^D x\,\left(|D_\mu\varphi|^2 + m^2\bar\varphi\varphi\right) \tag{4.55}$$

is invariant. The invariant action of the gauge field is

$$S_A = \frac{1}{4} \int \mathrm{d}^D x \, F_{\mu\nu}^{a\ 2} = -\frac{1}{2} \int \mathrm{d}^D x \, \mathrm{tr}[F_{\mu\nu}^2], \qquad (4.56)$$

where $F_{\mu\nu}$ is written using the matrices T^a of the fundamental representation. The gauge invariance of S_A is a consequence of (4.52) and the cyclicity of the trace. The theory (4.56) is called non-Abelian Yang-Mills theory. Note that it is an interacting theory.

The free-field limit $g \to 0$ of (4.56) describes a set of $N^2 - 1$ free photons. When we want to make sense of the functional integral of a non-Abelian gauge theory we meet the same problems we found in QED. The propagator can be defined only after a gauge fixing. To this purpose, a set of Faddeev-Popov ghosts C^a, together with their antighost partners $\bar{C}^a$ and the Lagrange multipliers B^a, have to be introduced, one for each parameter Λ^a of the local symmetry. Breaking the non-Abelian gauge symmetry in a way that does not affect the invariance of the physical quantities is more challenging than in QED. However, the gauge-fixing procedure becomes relatively simple, if we endow the symmetry with an appropriate canonical formalism.

Chapter 5

Canonical gauge formalism

The renormalization of gauge theories can be treated efficiently by means of a canonical formalism, known as Batalin-Vilkovisky formalism, which is equipped with proper notions of parentheses, canonical transformations, and other tools to make the key operations we need with a relatively small effort. The "time evolution" associated with it is just the gauge transformation. We do not need to explicitly introduce a "time" coordinate θ, because it would be an anticommuting parameter, so every function of θ has a Taylor expansion that stops at the first order.

Briefly, the Batalin-Vilkovisky formalism is a practical tool to (i) gather the key properties of the action, the infinitesimal symmetry transformations and their algebra into a single equation, (ii) fix the gauge and have control on the gauge fixing with a straightforward procedure, (iii) prove that the gauge theories can be renormalized by preserving gauge invariance to all orders, (iv) prove that the physical quantities are gauge independent, and (v) study the anomalies of the global and gauge symmetries to all orders. Combined with the dimensional regularization (or its modifications and upgrades), it allows us to derive these and several other properties in a systematic way, to all orders, and with much less effort than is required by the alternative approaches.

We generically refer to the resulting formal apparatus by calling it "canonical gauge formalism" for quantum field theory. It is well suited to study local symmetries, but also works for global symmetries.

In the rest of this chapter we mostly work at the bare level, but drop the subscripts ᴮ normally used to denote the bare quantities. The renormalization program will be carried out in detail in the next chapters. Among the other things, we will have to prove that the key formal structure is preserved by the subtraction of divergences.

We collect the classical fields into a single row

$$\phi^i = (A^a_\mu, \bar\psi, \psi, \varphi)$$

and assume that a classical action $S_c(\phi)$ is given, which is invariant under some infinitesimal transformations

$$\delta_\Lambda \phi^i = R^i_c(\phi, \Lambda), \tag{5.1}$$

that is to say,

$$\delta_\Lambda S_c = \int \delta_\Lambda \phi^i \frac{\delta_l S_c}{\delta \phi^i} = \int R^i_c(\phi, \Lambda) \frac{\delta_l S_c}{\delta \phi^i} = 0. \tag{5.2}$$

Here $\Lambda(x)$ denotes the local parameters of the symmetry.

5.1   General idea behind the canonical gauge formalism

We first introduce the basic ideas behind the canonical formalism, without paying too much attention to the details, such as the statistics of the fields, the correct relative positions of the fields and the sources, and some crucial minus signs that will be dealt with shortly after. It is useful to have a general idea of what we want to do, before plunging into the technical aspects. We go through the systematics in the next section.

The functions $R^i_c(\phi, \Lambda)$ are local, and linear in Λ. Apart from this, they can be arbitrary, possibly nonlinear in ϕ. This means that, once we get to the renormalization, they renormalize independently of the elementary fields. In other words, sooner or later we have to regard them as independent composite fields, so it is better to develop a formalism that does that from the start.

We know that the renormalization of composite fields can be studied by adding them to the action, coupled to external sources K. This defines the extended action

$$S^\Lambda_c(\phi, K) = S_c(\phi) - \int R^i_c(\phi, \Lambda) K_i.$$

The identity (5.2) can be written in the form

$$\delta_\Lambda S_c = (S_c^\Lambda, S_c) = \int \mathrm{d}^D x \left\{ \frac{\delta S_c^\Lambda}{\delta \phi^i(x)} \frac{\delta S_c}{\delta K_i(x)} - \frac{\delta S_c^\Lambda}{\delta K_i(x)} \frac{\delta S_c}{\delta \phi^i(x)} \right\} = 0. \quad (5.3)$$

This expression is appealing, because it reminds us of a canonical formalism, once the sources K are viewed as canonically conjugate to the elementary fields ϕ. The symmetry transformation of a functional $X(\phi)$ can be expressed as the parenthesis with S_c^Λ:

$$\delta_\Lambda X = (S_c^\Lambda, X) = \int R_c^i(\phi, \Lambda) \frac{\delta X}{\delta \phi^i}.$$

The identity (5.3), however, just tells us about the symmetry transformations, but does not incorporate the algebra of the transformations, in particular its closure. Closure means that the commutator $[\delta_\Lambda, \delta_\Sigma]$ of two transformations δ_Λ and δ_Σ, with parameters Λ and Σ, is a symmetry transformation $\delta_{\Delta(\Lambda, \Sigma)}$ of the same algebra, with certain parameters $\Delta(\Lambda, \Sigma)$:

$$[\delta_\Lambda, \delta_\Sigma] = \delta_{\Delta(\Lambda, \Sigma)}. \quad (5.4)$$

A priori, renormalization may affect both the action S_c and the transformations R_c^i, as well as the closure relations (5.4). Thus, it is important to collect these three pieces of information into a unique extended functional. By doing so, we build a powerful formalism to track how each ingredient renormalizes, and what role it plays inside the generating functionals.

In formula (5.3) we have two different functionals, S_c and S_c^Λ. Moreover, S_c, does not contain the sources K. As said, we would like to collect everything into a unique extended functional, and find an identity that involves only that functional. Now, the parenthesis $(S_c^\Lambda, S_c^\Lambda) = 0$ is trivial, because it is the subtraction of the term

$$\int \frac{\delta S_c^\Lambda}{\delta \phi^i} \frac{\delta S_c^\Lambda}{\delta K_i} \quad (5.5)$$

with itself. Nevertheless, the expression (5.5) goes into the right direction, as we can see if we split it into two pieces, the contributions that do not contain K and the contributions that are linear in K:

$$\int R_c^i \frac{\delta S_c}{\delta \phi^i}, \qquad \int R_c^i \frac{\delta R_c^j}{\delta \phi^i} K_j.$$

The former gives the transformation of the action S_c, and the latter somehow points to the transformation of the transformation, that is to say, the closure of the algebra.

A trick to make the terms (5.5) sum up, instead of canceling out, is to change the relative statistics of ϕ^i and K_i, and distinguish the left and the right derivatives. The resulting definition of parentheses, and the other details, are given below. We obtain a formalism that satisfies all the usual properties of a canonical formalism, once they are appropriately adapted, including a generalized Jacobi identity. Here we just anticipate that we get something like

$$(S_c^\Lambda, S_c^\Lambda) = 2 \int \frac{\delta S_c^\Lambda}{\delta \phi^i} \frac{\delta S_c^\Lambda}{\delta K_i}. \tag{5.6}$$

Again, this cannot be the final answer, because the "double" Λ transformation contained in (5.6) is not really a commutator. However, it becomes a commutator once we also play with the statistics of Λ, and provide suitable transformation rules for the Λ's themselves.

To study the closure, we need two independent parameters, e.g., Λ and Σ, so S_c^Λ is certainly inadequate to contain the closure relations (5.4). On a functional X, we should have

$$[\delta_\Lambda, \delta_\Sigma] X = (S_c^\Lambda, (S_c^\Sigma, X)) - (S_c^\Sigma, (S_c^\Lambda, X)) = \delta_{\Delta(\Lambda, \Sigma)} X = (S_c^{\Delta(\Lambda, \Sigma)}, X). \tag{5.7}$$

We may expect that the generalized Jacobi identity for the parentheses allows us to replace the expression between the first and second equal signs of (5.7) with something like $((S_c^\Lambda, S_c^\Sigma), X)$. Then (5.7) should give

$$((S_c^\Lambda, S_c^\Sigma) - S_c^{\Delta(\Lambda, \Sigma)}, X) = 0, \tag{5.8}$$

for every X, so we can express the closure by means of a relation of the form

$$(S_c^\Lambda, S_c^\Sigma) = S_c^{\Delta(\Lambda, \Sigma)}.$$

We still have three functionals, instead of one.

It turns out that we can collect everything into a single functional, if we replace $\Lambda(x)$ and $\Sigma(x)$ by $\theta C(x)$ and $\theta' C(x)$, respectively, where θ and θ' are anticommuting parameters (which we can drop after moving them to the right, or left, of each identity) and $C(x)$ is an anticommuting field, to be identified with the Fadeev-Popov ghosts. The main virtue of such replacements is that they can generate commutators quite easily. For example, if we set $C^a = \theta \Lambda^a + \theta' \Sigma^b$, we obtain

$$C^a C^b = \theta \theta' (\Lambda^a \Sigma^b - \Sigma^a \Lambda^b).$$

These tricks allow us to work with a unique, but anticommuting, $C(x)$ and a unique extended action. If done properly, the operations encoded into replacements like $\Lambda \to \theta C$, $\Sigma \to \theta' C$, $C \to \theta \Lambda + \theta' \Sigma$ are completely reversible, so they do not cause any loss of information. Finally, $\Delta(C, C)$ is identified with the transformation of C itself, apart from a proportionality factor.

The new extended action is something of the form

$$S'_c(\phi, C, K) = S_c(\phi) + \int \left(K_i R^i_c(\phi, C) - \frac{1}{2} K_C \Delta(C, C) \right),$$

where K_C are sources for the C transformations. Next, we have an identity like

$$(S'_c, S'_c) = 2 \int \frac{\delta S'_c}{\delta \phi^i} \frac{\delta S'_c}{\delta K_i} + 2 \int \frac{\delta S'_c}{\delta C} \frac{\delta S'_c}{\delta K_C} = 0. \tag{5.9}$$

The terms proportional to K_i in this expression give the closure of the algebra. The terms proportional to K_C cancel out by themselves, because they are just the Jacobi identity of the Lie algebra.

Summarizing, once the crucial identity $(S'_c, S'_c) = 0$ is satisfied, the extended action S'_c incorporates the invariant action, the symmetry transformations of the fields, the closure of the algebra and its Jacobi identity.

5.2 Systematics of the canonical gauge formalism

Without further premises, we are ready to present the systematics of the canonical gauge formalism. Make the substitution $\Lambda(x) \to \theta C(x)$ in the identity (5.2), then move θ to the far left and drop it. Since R^i is linear in Λ, we get an identity of the form

$$\int R^i(\phi, C) \frac{\delta_l S_c}{\delta \phi^i} = 0. \tag{5.10}$$

The functions $R^i(\phi, C)$ are such that

$$R^i_c(\phi, \theta C) = \theta R^i(\phi, C) \tag{5.11}$$

and may differ from $R_c^i(\phi, C)$ by a sign, depending on the statistics of ϕ.

The fields C are denoted this way, because they coincide with the Faddeev-Popov ghosts already met. For the moment, we do not need to introduce the antighosts $\bar{C}$ and the Lagrange multipliers B. They are useful to fix the gauge, but they are not basic ingredients of the canonical formalism.

For definiteness, we work in non-Abelian Yang-Mills theory coupled to fermions and scalar fields. QED can be seen as a particular case with gauge group $G = U(1)$. We include the fields and the ghosts into the extended row

$$\Phi^\alpha = (A_\mu^a, C^a, \bar{\psi}, \psi, \varphi).$$

The conjugate row made by the sources is

$$K_\alpha = (K_a^\mu, K_C^a, K_{\bar{\psi}}, K_\psi, K_\varphi).$$

We define the statistics ε_Φ, ε_K, ε_λ, ε_X of a field Φ, a source K, a parameter λ or a functional X to be zero if the field, source, parameter or functional is bosonic, one if it is fermionic. We define the statistics of the sources as opposite to the statistics of the fields that are conjugate to them:

$$\varepsilon_{K_\alpha} = \varepsilon_{\Phi^\alpha} + 1 \bmod 2. \tag{5.12}$$

Given two functionals $X(\Phi, K)$ and $Y(\Phi, K)$ of the fields and the sources, we define their *antiparentheses* as the functional

$$(X, Y) \equiv \int \mathrm{d}^D x \left\{ \frac{\delta_r X}{\delta \Phi^\alpha(x)} \frac{\delta_l Y}{\delta K_\alpha(x)} - \frac{\delta_r X}{\delta K_\alpha(x)} \frac{\delta_l Y}{\delta \Phi^\alpha(x)} \right\}, \tag{5.13}$$

where the sum over α is understood, and δ_r, δ_l denote the left and right functional derivatives, respectively. Observe that if X and Y are local functionals, then (X, Y) is a local functional.

The antiparentheses satisfy the properties

$$(Y, X) = -(-1)^{(\varepsilon_X + 1)(\varepsilon_Y + 1)}(X, Y),$$
$$(-1)^{(\varepsilon_X + 1)(\varepsilon_Z + 1)}(X, (Y, Z)) + \text{cyclic permutations} = 0, \tag{5.14}$$

and $\varepsilon_{(X,Y)} = \varepsilon_X + \varepsilon_Y + 1$, which can be verified straightforwardly. In particular, formula (5.14) is the Jacobi identity. Immediate consequences are

$$(F, F) = 0, \qquad (B, B) = 2 \int \frac{\delta_r B}{\delta \Phi^\alpha} \frac{\delta_l B}{\delta K_\alpha} = -2 \int \frac{\delta_r B}{\delta K_\alpha} \frac{\delta_l B}{\delta \Phi^\alpha}, \tag{5.15}$$

if the functionals F and B have fermionic and bosonic statistics, respectively. In (5.15), as often below, we understand the integrations over the spacetime points associated with the repeated indices α, β,.... Another important consequence is

$$(X, (X, X)) = 0 \qquad (5.16)$$

for every functional X. This property follows from the Jacobi identity (5.14), and is useful to study the anomalies.

The action $S(\Phi, K)$ is defined as the solution of the *master equation*

$$(S, S) = 0, \qquad (5.17)$$

with the *boundary conditions*

$$S(\Phi, 0) = S_c(\phi), \qquad -\frac{\delta_r S(\Phi, K)}{\delta K_i}\bigg|_{K=0} = R^i(\phi, C). \qquad (5.18)$$

In the naïve derivation given above, the extended action S'_c was linear in the sources K. This is actually true in all the applications we have in mind, at least at the tree level. Thus, we write the general solution of the master equation in the form

$$S(\Phi, K) = S_c(\phi) - \int R^\alpha(\Phi) K_\alpha = S_c(\phi) - \int R^i(\phi, C) K_i - \int R^a_C(\Phi) K^a_C, \qquad (5.19)$$

where the functions $R^a_C(\Phi)$, related somehow to $\Delta(\Lambda, \Lambda)$, have yet to be determined. The signs have been adjusted to match the choices of statistics we have made. It can be shown that the linearity of $S(\Phi, K)$ in K means that the gauge algebra closes off shell.

More explicitly, using the last expression of (5.15), we find the formula

$$0 = (S, S) = 2 \int R^\alpha(\Phi) \frac{\delta_l S}{\delta \Phi^\alpha} = 2 \int \left[R^i(\phi, C) \frac{\delta_l S}{\delta \phi^i} + R^a_C(\Phi) \frac{\delta_l S}{\delta C^a} \right].$$

The terms of order 0 in K are twice the identity (5.10), while the terms of order 1 in K give the formula

$$0 = -2 \int R^\alpha(\Phi) \frac{\delta_l}{\delta \Phi^\alpha} \int R^\beta(\Phi) K_\beta,$$

which implies

$$0 = \int R^\alpha(\Phi) \frac{\delta_l R^\beta(\Phi)}{\delta \Phi^\alpha} \qquad (5.20)$$

for every β. Taking $\beta = i$, we find

$$0 = \int R^j(\phi, C)\frac{\delta_l R^i(\phi, C)}{\delta\phi^j} + \int R^a_C(\Phi)\frac{\delta_l R^i(\phi, C)}{\delta C^a}. \tag{5.21}$$

Since the functions R^i are linear in C, and the functions R^a_C have bosonic statistics, the last term is equal to $R^i_c(\phi, R_C(\Phi))$. Setting $C^a = \theta\Lambda^a + \theta'\Sigma^a$ in (5.21), where θ and θ' are anticommuting parameters, we obtain

$$0 = \int R^j(\phi, \theta\Lambda^a + \theta'\Sigma)\frac{\delta_l R^i(\phi, \theta\Lambda^a + \theta'\Sigma)}{\delta\phi^j} + R^i_c(\phi, R^a_C|_{C\to\theta\Lambda+\theta'\Sigma}). \tag{5.22}$$

To read the content of this formula, we need to move θ and θ' to the left and drop them. To achieve this goal, first note that setting $C^a = \theta'\Lambda^a$ in formula (5.11), dropping θ to the left, and redefining θ' as θ, we also obtain

$$\theta R^i_c(\phi, \Lambda^a) = R^i(\phi, \theta\Lambda^a). \tag{5.23}$$

Moreover, the functions R^a_C must be quadratic in the ghosts C. Then, $R^a_C|_{C\to\theta\Lambda+\theta'\Sigma}$ is proportional to $\theta\theta'$. Let us write

$$R^a_C(\Phi)|_{C\to\theta\Lambda+\theta'\Sigma} = -\theta\theta'\Delta(\Lambda, \Sigma). \tag{5.24}$$

This formula can be taken as the very definition of $R^a_C(\Phi)$, where the functions $\Delta(\Lambda, \Sigma)$ are assumed to be known from the closure relation (5.4). In the end, the expressions $R^a_C(\Phi)$ just depend on C, since in gauge theories (and gravity) the quantities $\Delta(\Lambda, \Sigma)$ are ϕ independent.

Using (5.23) and (5.24), formula (5.22) gives

$$\int \theta R^j_c(\phi, \Lambda^a)\frac{\delta_l \theta' R^i_c(\phi, \Sigma)}{\delta\phi^j} + \int \theta' R^j_c(\phi, \Sigma)\frac{\delta_l \theta R^i_c(\phi, \Lambda^a)}{\delta\phi^j} = \theta\theta' R^i_c(\phi, \Delta(\Lambda, \Sigma)). \tag{5.25}$$

Moving θ and θ' to the left, and using (5.1), we obtain

$$\theta\theta' \int \left(\delta_\Lambda\phi^j\frac{\delta_l(\delta_\Sigma\phi^i)}{\delta\phi^j} - \delta_\Sigma\phi^j\frac{\delta_l(\delta_\Lambda\phi^i)}{\delta\phi^j}\right) = \theta\theta' R^i_c(\phi, \Delta(\Lambda, \Sigma)),$$

or

$$\theta\theta'[\delta_\Lambda, \delta_\Sigma]\phi^i = \theta\theta'\delta_{\Delta(\Lambda,\Sigma)}\phi^i,$$

which is equivalent to (5.4).

Finally, taking $R^\beta \to R^a_C$ in (5.20), we get

$$0 = \int R^b_C(\Phi) \frac{\delta_l R^a_C(\Phi)}{\delta C^b}, \tag{5.26}$$

having used that the functions $R^a_C(\Phi)$ just depend on C. Formula (5.26), which is the closure of the closure, in some sense, is just the Jacobi identity of the lie algebra.

For example, in non-Abelian Yang-Mills theories, we have (on fermions ψ, for definiteness) $\delta_\Lambda \psi^i = -g T^a_{ij} \Lambda^a \psi^j$, so

$$[\delta_\Lambda, \delta_\Sigma] \psi^i = g^2 [T^a, T^b]_{ij} \psi^j \Sigma^a \Lambda^b = -g^2 T^c_{ij} \psi^j f^{abc} \Lambda^a \Sigma^b = \delta_{\Delta(\Lambda, \Sigma)} \psi^i,$$

hence

$$\Delta^a(\Lambda, \Sigma) = g f^{abc} \Lambda^b \Sigma^c.$$

Using this expression in (5.24), we find

$$R^a_C(\Phi)|_{C \to \theta\Lambda + \theta'\Sigma} = -\theta\theta' g f^{abc} \Lambda^b \Sigma^c = -\frac{g}{2} g f^{abc} (\theta\Lambda + \theta'\Sigma)^b (\theta\Lambda + \theta'\Sigma)^c,$$

whence

$$R^a_C(\Phi) = -\frac{g}{2} f^{abc} C^b C^c. \tag{5.27}$$

Thus, the identity (5.26) gives

$$0 = f^{abc} f^{bde} C^c C^d C^e.$$

Since the C's are anticommuting, this equation is equivalent to (4.31).

In the Abelian case, $\Delta(\Lambda, \Sigma) = 0$, so $R_C(\Phi) = 0$.

The solution (5.19) to the master equation is called *minimal*, because it contains the minimal set of fields. The minimal solution is not sufficient to gauge-fix the theory and define the propagators of the gauge fields, because it does not contain the antighosts and the Lagrange multipliers. We can include them by enlarging the sets of fields and sources to

$$\Phi^\alpha = (A^a_\mu, C^a, \bar{C}^a, B^a, \bar{\psi}, \psi, \varphi), \quad K_\alpha = (K^\mu_a, K^a_C, K^a_{\bar{C}}, K^a_B, K_{\bar{\psi}}, K_\psi, K_\varphi). \tag{5.28}$$

Again, the statistics of the sources are defined to be opposite to those of their conjugate fields.

It is easy to prove that if $S_{\min}(\Phi, K)$ is a minimal solution to the master equation the extended action

$$S_{\min}(\Phi, K) - \int B^a K^a_{\bar{C}}$$

is also a solution. We call it the *extended solution* to the master equation. This extension is sufficient for the purposes of gauge fixing. From now we understand that the sets of fields and sources are (5.28) and the general solution to the master equation is

$$S(\Phi, K) = S_c(\phi) - \int \left[R^i(\phi, C) K_i + R^a_C(\Phi) K^a_C + B^a K^a_{\bar{C}} \right]. \qquad (5.29)$$

It is also useful to introduce the *ghost number*,

$$\text{gh}(A) = \text{gh}(\psi) = \text{gh}(\bar{\psi}) = \text{gh}(\varphi) = \text{gh}(B) = 0, \quad \text{gh}(C) = 1, \quad \text{gh}(\bar{C}) = -1,$$

together with $\text{gh}(K_\alpha) = -\text{gh}(\Phi_\alpha) - 1$. Indeed, the global $U(1)$ transformation

$$\Phi^\alpha \to \Phi^\alpha e^{i\sigma\text{gh}(\Phi^\alpha)}, \qquad K_\alpha \to K_\alpha e^{i\sigma\text{gh}(K_\alpha)}, \qquad (5.30)$$

σ being a constant parameter, is a symmetry of the actions we are going to work with, as well as the functional integration measure. The ghost number is trivially preserved by the Feynman rules and the diagrammatics, and so by the radiative corrections and renormalization.

The fermionic number of a field or a source is equal to zero or one, depending on whether the field or source is a boson or a fermion. The statistics of a field or a source is equal to the sum of its fermionic number plus its ghost number, modulo 2. For example, the sources K_ψ and $K_{\bar{\psi}}$ associated with the Dirac fermions are commuting objects, since they are fermions, but they also have odd ghost numbers. Thus, K_ψ and $K_{\bar{\psi}}$ are "fermions with bosonic statistics", while C and $\bar{C}$ are "bosons with fermionic statistics".

Now we are ready to derive the solutions of the master equation for Abelian and non-Abelian gauge theories. In quantum electrodynamics formulas (4.7) give the functions $R^i_c(\phi, \Lambda)$. Replacing Λ by θC and using $R^i_c(\phi, \theta C) = \theta R^i(\phi, C)$, we obtain the functions $R^i(\phi, C)$ for A_μ, ψ and $\bar{\psi}$, which read

$$\partial_\mu C, \qquad -ieC\psi, \qquad -ie\bar{\psi}C,$$

respectively. The functions R_C^a associated with the ghosts can be derived from the closure of the algebra. Since it is trivial in the Abelian case, we just have $R_C^a = 0$. Thus, the extended solution of the master equation reads

$$S(\Phi, K) = S_c(\phi) - \int \left(\partial_\mu C K_\mu - ie\bar\psi C K_{\bar\psi} - ieK_\psi C\psi + BK_{\bar C} \right), \quad (5.31)$$

where the classical action is

$$S_c(\phi) = \frac{1}{4} \int F_{\mu\nu}^2 + \int \bar\psi (\slashed\partial + ie\slashed A + m)\psi.$$

It is easy to check that (5.17) is indeed satisfied.

In non-Abelian Yang-Mills theory we start from (4.53) to read the functions $R_c^i(\phi, \Lambda)$, then replace Λ by θC and use $R_c^i(\phi, \theta C) = \theta R^i(\phi, C)$ again. We find that

$$\partial_\mu C^a + g f^{abc} A_\mu^b C^c, \qquad -g T_{ij}^a C^a \psi^j, \qquad -g \bar\psi^j T_{ij}^a C^a, \qquad (5.32)$$

are the functions $R^i(\phi, C)$ for the gauge potential A_μ^a, the fermions ψ^i in the fundamental representation and their conjugates $\bar\psi^i$, respectively. The functions $R_C^a(\Phi)$ are given by (5.27).

The solution reads

$$S(\Phi, K) = S_c(\phi) + g \int \left(\bar\psi^i T_{ij}^a C^a K_{\bar\psi}^j + K_\psi^i T_{ij}^a C^a \psi^j \right)$$

$$- \int \left[(\partial_\mu C^a + g f^{abc} A_\mu^b C^c) K_\mu^a - \frac{g}{2} f^{abc} C^b C^c K_C^a + B^a K_{\bar C}^a \right], \quad (5.33)$$

where the classical action is

$$S_c(\phi) = \frac{1}{4} \int F_{\mu\nu}^{a\,2} + \int \bar\psi^i (\delta_{ij}\slashed\partial + g T_{ij}^a \slashed A^a + m\delta_{ij})\psi^j. \quad (5.34)$$

In the next chapters we prove the renormalizability of both theories.

From (5.33) and (5.34) we can read the dimensions [ ] of the fields and the sources, as well as their statistics ε_{Φ^A}, ε_{K_A}. The ghost numbers $\mathrm{gh}(K)$ of the sources are obtained by demanding that (5.33) be invariant under (5.30). We have the tables

	A_μ^a	C^a	$\bar C^a$	B^a	$\bar\psi$	ψ	φ	
[ ]	$\frac{D}{2} - 1$	$\frac{D}{2} - 1$	$\frac{D}{2} - 1$	$\frac{D}{2}$	$\frac{D-1}{2}$	$\frac{D-1}{2}$	$\frac{D}{2} - 1$	(5.35)
gh	0	1	-1	0	0	0	0	

	K_a^μ	K_C^a	$K_{\bar{C}}^a$	K_B^a	$K_{\bar{\psi}}$	K_ψ	K_φ
[ ]	$\frac{D}{2}$	$\frac{D}{2}$	$\frac{D}{2}$	$\frac{D}{2}-1$	$\frac{D-1}{2}$	$\frac{D-1}{2}$	$\frac{D}{2}$
gh	-1	-2	0	-1	-1	-1	-1

(5.36)

Now we stress a property that will be useful later on, to prove the renormalizability of non-Abelian Yang-Mills theories. Theorem 5 allows us to work without the matrices T^a and $\mathcal{T}^a$, and just use the invariant tensors (4.42) and the gauge-field variables (4.51). Distinguishing upper and lower indices, we have $A^i_{\mu\,j} = -iT^{ai}_{\ \ j}A^a_\mu$. Using (4.32), the converse formula reads $A^a_\mu = -2iT^{aj}_{\ \ i}A^i_{\mu\,j}$. Similarly, the ghosts, antighosts and Lagrange multipliers can be written as $C^i_j = -iT^{ai}_{\ \ j}C^a$, $\bar{C}^i_j = -iT^{ai}_{\ \ j}\bar{C}^a$, and $B^i_j = -iT^{ai}_{\ \ j}B^a$. The sources must be defined a little bit differently, to have a canonical transformation (see below): $K^i_{\mu\,j} = -2iT^{ai}_{\ \ j}K^a_\mu$, $K^a_\mu = -iT^{aj}_{\ \ i}K^i_{\mu\,j}$, etc.

When the fermions are in the fundamental representation, the solution of the master equation is

$$S(\Phi, K) = S_c(\phi) - \int \left(\partial_\mu C^i_j + ig(A^i_{\mu k}C^k_j - A^k_{\mu j}C^i_k) \right) K^j_{\mu i}$$

$$+ig \int \left(\bar{\psi}_i C^i_j K^j_{\bar\psi} + K_{\psi i} C^i_j \psi^j \right) + ig \int C^i_k C^k_j K^j_{Ci} - \int B^i_j K^j_{\bar{C}i}, \quad (5.37)$$

where

$$S_c(\phi) = \frac{1}{4}\int (F^i_{\mu\nu j})^2 + \int \bar{\psi}_i(\delta^i_j \slashed{\partial} + ig\slashed{A}^i_{\ j} + m\delta^i_j)\psi^j, \quad (5.38)$$

and

$$F^i_{\mu\nu j} = \partial_\mu A^i_{\nu j} - \partial_\nu A^i_{\mu j} + ig(A^i_{\mu k}A^k_{\nu j} - A^i_{\nu k}A^k_{\mu j}).$$

To derive these expressions we have also used formula (4.46).

A matter field ψ^I in an irreducible representation r can be denoted by $\psi^{i_1\cdots i_n}_{j_1\cdots j_m}$, if the indices have appropriate symmetry properties. From (4.41) we derive that its contribution to the solution $S(\Phi, K)$ proportional to the sources K becomes

$$ig \int \left(\psi^{i_1\cdots i_n}_{k_1 j_2\cdots j_m} C^{k_1}_{j_1} + \cdots + \psi^{i_1\cdots i_n}_{j_1\cdots j_{m-1}k_m} C^{k_m}_{j_m} \right) K^{j_1\cdots j_m}_{i_1\cdots i_n}$$

$$-ig \int \left(\psi^{l_1 i_2\cdots i_n}_{j_1\cdots j_m} C^{i_1}_{l_1} + \cdots + \psi^{i_1\cdots i_{n-1}l_n}_{j_1\cdots j_m} C^{i_n}_{l_n} \right) K^{j_1\cdots j_m}_{i_1\cdots i_n}. \quad (5.39)$$

Finally, a generic vertex has the form

$$\phi^{i_1\cdots i_n}_{j_1\cdots j_m} \psi^{k_1\cdots k_p}_{l_1\cdots l_q} \chi^{u_t\cdots u_r}_{v_1\cdots v_s} \cdots \quad (5.40)$$

with indices contracted by means of the invariant tensors (4.42).

An important theorem states that Yang-Mills theory in practice exhausts the gauge theories of vector fields.

Theorem 7 *The most general local, power counting renormalizable quantum field theory of vector fields is a Yang-Mills theory based on a Lie algebra.*

Proof. To prove this theorem we can take advantage of the canonical formalism, because we know that it collects the properties of the gauge algebra in a compact form. Let A_μ^I denote the set of gauge vectors contained in the theory. In the free-field limit, the theory must be invariant under the Abelian symmetry $\delta A_\mu^I = \partial_\mu \Lambda^I$. Writing $\Lambda^I = \theta C^I$, as usual, the functions $R_\mu^I(\phi, C)$ that encode the complete symmetry transformations must be equal to $\partial_\mu C^I$ plus interaction terms. By locality, ghost number conservation and power counting, the most general source sector of the minimal solution to the master equation must have the form

$$-(\partial_\mu C^I + A_\mu^J C^K \kappa^{JKI}) K_\mu^I + \frac{1}{2} C^J C^K h^{JKI} K_C^I,$$

where κ^{IJK} and h^{JKI} are numerical constants and the h^{JKI} are antisymmetric in J and K. Now we study the constraints imposed by the master equation $(S, S) = 0$. It is easy to show that the terms proportional to K_C^I in $(S, S) = 0$ imply that the constants h^{JKI} satisfy the Jacobi identity

$$h^{IJK} h^{KLM} + h^{LIK} h^{KJM} + h^{JLK} h^{KIM} = 0.$$

Since both assumptions (4.30) and (4.31) are satisfied, the constants h^{JKI} define a Lie algebra. It is also straightforward to check that the terms proportional to K^I in $(S, S) = 0$ give $\kappa^{IJK} = h^{IJK}$. Thus, the gauge transformations have the Yang-Mills form

$$\delta_\Lambda A_\mu^I = \partial_\mu \Lambda^I + A_\mu^J \Lambda^K h^{JKI},$$

which proves the theorem. $\square$

We stress again that we have not proved the renormalizability, yet, but this theorem anticipates that if Yang-Mills theory is renormalizable, it is unique.

Given a functional Y, we can view the antiparenthesis (Y, X) as a map acting on the space of functionals X. Choosing $Y = S$, the map (S, X) is nilpotent, because of the Jacobi identity (5.14) and the master equation (5.17). Indeed,

$$(S, (S, X)) = \frac{1}{2}((S, S), X) = 0. \tag{5.41}$$

On the fields and the sources, we have

$$(S, \Phi^\alpha) = R^\alpha(\Phi), \qquad (S, K_\alpha) = \frac{\delta_r S}{\delta \Phi^\alpha}.$$

Basically, (S, K_α) is the Φ^α field equation, plus $\mathcal{O}(K)$.

The map (S, X) sends functionals $\mathcal{G}(\Phi)$ that depend only on the fields into functionals that depend only on the fields:

$$(S, \mathcal{G}(\Phi)) = \int R^\alpha(\Phi) \frac{\delta_l \mathcal{G}(\Phi)}{\delta \Phi^\alpha}.$$

On the functionals $\mathcal{G}(\phi)$ that depend only on the physical fields ϕ, the map is precisely the gauge transformation:

$$(S, \mathcal{G}(\phi)) = \int R^i(\phi, C) \frac{\delta_l \mathcal{G}(\phi)}{\delta \phi^i}.$$

In particular, $(S, S_c(\phi)) = 0$, which is nothing but the gauge invariance of the classical action $S_c(\phi)$.

It is always possible to generate new solutions $S(\Phi, K)$ of the master equation, which are possibly nonlinear in K, by means of field and source redefinitions that preserve the master equation itself (or the antiparentheses, in which case they are canonical transformations, see below). However, not all the solutions that are nonlinear in K can be obtained this way. To understand this issue better, consider again the relation

$$0 = (S, S) = -2 \int \frac{\delta_r S}{\delta K_\alpha} \frac{\delta_l S}{\delta \Phi^\alpha}. \tag{5.42}$$

The K-independent contributions are always the identity (5.10), but when $S(\Phi, K)$ is not linear in K, the terms of (5.42) that are linear in K (which encode the closure of the algebra) contain extra contributions proportional to the field equations. If there exists no canonical transformation that absorbs the extra terms away, it means that the gauge algebra does not close off shell, but just on shell. Still, it may be possible to close it off shell by adding a finite number of extra fields. In this book we mainly focus on such cases, because the symmetry algebras that close off shell are those that play a major role in physical applications. Nevertheless, it is important to know that more general structures may exist, and that the solution (5.19) depends on the field variables we choose.

The canonical formalism does not apply only to local functionals, such as the action S, but also to nonlocal functionals, such as the effective action Γ. For this reason, it is necessary to prove some general properties before proceeding. We have remarked above that the antiparentheses map local functionals X and Y into a local functional (X, Y). We now prove that they also map one-particle irreducible functionals X and Y into a one-particle irreducible functional (X, Y). Define the operator

$$\mathcal{V} \equiv \int \mathrm{d}^D x \frac{\overleftarrow{\delta_r}}{\delta \Phi^i(x)} \frac{\overrightarrow{\delta_l}}{\delta K_i(x)}.$$

We focus on the contribution $X\mathcal{V}Y$ to (X, Y) in (5.13), since the other contribution can be treated in an analogous way. Note that if X and Y are one-particle irreducible, a functional derivative with respect to $\Phi^\alpha(x)$ is an amputated Φ^α leg, and a functional derivative with respect to $K_\alpha(x)$ is an insertion of $R^\alpha(\Phi(x))$. In particular, no propagators are attached to such legs. The operator $\mathcal{V}$ generates a sort of new vertex, whose legs are the legs attached to $\Phi(x)$ in $\delta_r X / \delta \Phi(x)$, plus the legs attached to $K(x)$ in $\delta_l Y / \delta K(x)$. Since the diagrams of $\delta_r X / \delta \Phi(x)$ and $\delta_l Y / \delta K(x)$ are irreducible, the contribution $X\mathcal{V}Y$ to (X, Y) is also irreducible. Diagrammatically, we have

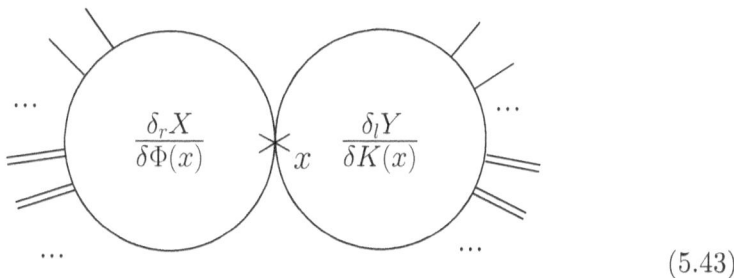

$$(5.43)$$

where the double lines are sources and the single lines are fields.

5.3   Canonical transformations

A canonical transformation of the fields and the sources is a transformation

$$\Phi^{\alpha\prime}(\Phi, K), \qquad K'_\alpha(\Phi, K),$$

that preserves the antiparentheses, that is to say such that

$$(X', Y')' = (X, Y)$$

for every pairs of functionals X and Y, where X' and Y' are defined as

$$X'(\Phi', K') = X(\Phi(\Phi', K'), K(\Phi', K')),$$
$$Y'(\Phi', K') = Y(\Phi(\Phi', K'), K(\Phi', K')),$$

and $(.,.)'$ denotes the antiparentheses calculated with respect to Φ' and K'.

A canonical transformation is generated by a functional $\mathcal{F}(\Phi, K')$. It can be expressed as

$$\Phi^{\alpha\prime} = \frac{\delta \mathcal{F}}{\delta K'_\alpha}, \qquad K_\alpha = \frac{\delta \mathcal{F}}{\delta \Phi^\alpha}. \tag{5.44}$$

Formula (5.12) implies that $\mathcal{F}$ is a functional of fermionic statistics. For this reason, there is no need to specify whether the derivatives of (5.44) are left or right.

The generating functional of the identity transformation is

$$\mathcal{I}(\Phi, K') = \int \mathrm{d}^D x\, \Phi^\alpha(x) K'_\alpha(x).$$

Let us inspect the most general canonical transformation, to understand what it can be useful for. We can write the generating functional as a sum of a term that is independent of the sources plus the rest:

$$\mathcal{F}(\Phi, K') = \Psi(\Phi) + \int K'_\alpha U^\alpha(\Phi, K'). \tag{5.45}$$

Then (5.44) gives

$$\Phi^{\alpha\prime} = U^\alpha(\Phi, K') + \int K'_\beta \frac{\delta_r U^\beta}{\delta K'_a}, \qquad K_\alpha = \frac{\delta \Psi(\Phi)}{\delta \Phi^\alpha} + \int K'_\beta \frac{\delta_r U^\beta}{\delta \Phi^A}. \tag{5.46}$$

Recall that we must eventually set the sources K to zero, since they are introduced just as tools to have control on the gauge symmetry. To illustrate the meaning of (5.45), we set them to zero after the canonical transformation. If we drop the terms proportional to K' in (5.46) we obtain

$$\Phi^{\alpha\prime} = U^\alpha(\Phi, 0), \qquad K_\alpha = \frac{\delta\Psi(\Phi)}{\delta\Phi^\alpha}. \tag{5.47}$$

The Φ transformation is a field redefinition. Instead, as we will explain later, the K transformation is a gauge fixing, or a change of gauge fixing. The K-dependent terms contained in (5.46) do not have a particular meaning. They are there to promote (5.47) to a canonical transformation, which is much easier to manipulate, because it preserves the antiparentheses. Thus,

Proposition 8 *the most general canonical transformation (5.45) is the combination of the most general field redefinition and the most general gauge fixing.*

Of course, the physics cannot depend on the field variables and gauge-fixing we use. Observe that if X is such that $(S, X) = 0$, then X' is such that $(S', X')' = 0$. In particular, the gauge symmetry transforms so as to keep a gauge invariant functional invariant.

5.4   Gauge fixing

We have gauge-fixed quantum electrodynamics in chapter 4. Now we are ready to gauge-fix the non-Abelian Yang-Mills theories. Denote the gauge-fixing functions of our choice with $\mathcal{G}^a(A)$. For example, $\mathcal{G}^a(A) = \partial_\mu A^a_\mu$ in the Lorenz gauge, and $\mathcal{G}^a(A) = -\nabla \cdot \mathbf{A}^a$ in the Coulomb gauge. Define the *gauge fermion*

$$\Psi(\Phi) = \int \bar{C}^a \left(-\frac{\lambda}{2} B^a + \mathcal{G}^a(A) \right). \tag{5.48}$$

The gauge fermion is a local functional of fermionic statistics that fixes the gauge in the way explained below. Its typical form is (5.48), but more general functionals can be chosen.

Working in the Lorenz gauge, for definiteness, define the gauge-fixed action as

$$S_\Psi(\Phi, K) = S(\Phi, K) + (S, \Psi). \tag{5.49}$$

It is easy to prove that S_Ψ and S are related by the canonical transformation generated by the functional

$$\mathcal{F}(\Phi, K') = \int \Phi^\alpha K'_\alpha + \Psi(\Phi). \tag{5.50}$$

Indeed, (5.44) gives

$$\Phi^{\alpha\prime} = \Phi^\alpha, \qquad K'_\alpha = K_\alpha - \frac{\delta\Psi(\Phi)}{\delta\Phi^\alpha}.$$

Recalling that the action is linear in the sources, we have from (5.19)

$$S(\Phi', K') = S(\Phi, K) + \int R^\alpha(\Phi)\frac{\delta\Psi(\Phi)}{\delta\Phi^\alpha} = S(\Phi, K) + (S, \Psi) = S_\Psi(\Phi, K).$$

Moreover,

Proposition 9 *If S satisfies the master equation, then every $S_\Psi = S + (S, \Psi(\Phi))$ satisfies the master equation.*

The reason is that the canonical transformations preserve the antiparentheses, so $(S, S) = 0$ if and only if $(S_\Psi, S_\Psi) = 0$. In particular, the gauge-fixing procedure preserves the master equation.

Working out S_Ψ explicitly, we find

$$S_\Psi(\Phi, K) = S_c(\phi) + S_{\text{gf}}(\Phi) - \int R^\alpha K_\alpha, \tag{5.51}$$

where

$$S_c(\phi) + S_{\text{gf}}(\Phi) = \int d^D x \left(\frac{1}{4} F_{\mu\nu}^{a\ 2} - \frac{\lambda}{2}(B^a)^2 + B^a \partial \cdot A^a - \bar{C}^a \partial_\mu D_\mu C^a \right). \tag{5.52}$$

Observe that the ghosts do not decouple.

The gauge-field propagator can be worked out from the free subsector of (5.52), after integrating B^a out, which gives the equivalent gauge-fixed action

$$S = \int d^D x \left(\frac{1}{4} F_{\mu\nu}^{a\ 2} + \frac{1}{2\lambda} (\partial_\mu A_\mu^a)^2 - \bar{C}^a \partial_\mu D_\mu C^a \right). \tag{5.53}$$

The result is

$$\langle A_\mu^a(k) A_\nu^b(-k) \rangle = \frac{\delta^{ab}}{k^2} \left(\delta_{\mu\nu} + (\lambda - 1)\frac{k_\mu k_\nu}{k^2} \right). \tag{5.54}$$

The ghost propagator is

$$\langle C^a(k)\bar{C}^b(-k)\rangle = \frac{\delta^{ab}}{k^2}. \tag{5.55}$$

Repeating the argument that leads to (4.24) we can check, in the Coulomb gauge, that the physical degrees of freedom are $2 \dim G$, as it must be.

The argument just given does not change when we add matter fields, since they are not interested by the gauge-fixing procedure. Clearly, in QED we get back (4.18) and (4.19).

Exercise 22 *Show that the action* $S_\Psi(\Phi, K)$ *continues to satisfy the master equation after integrating out the Lagrange multipliers* B^a.

Solution. Integrating B out is equivalent to replacing B with the solution of its own field equation, that is to say, making the replacement

$$B^a \to \frac{1}{\lambda}(\partial \cdot A^a - K^a_{\bar{C}}).$$

Then $S_\Psi(\Phi, K)$ becomes

$$\bar{S}_\Psi(\Phi, K) = S_\Psi + \frac{1}{2\lambda}\int (\lambda B^a + K^a_{\bar{C}} - \partial \cdot A^a)^2 = \frac{1}{4}\int F^{a\,2}_{\mu\nu} - \int \bar{C}^a \partial_\mu D_\mu C^a$$

$$+ \frac{1}{2\lambda}\int (K^a_{\bar{C}} - \partial \cdot A^a)^2 - \int (D_\mu C^a)K^a_\mu + \frac{g}{2}\int f^{abc}C^b C^c K^a_{\bar{C}}.$$

At this point, it is straightforward to check the master equation $(\bar{S}_\Psi, \bar{S}_\Psi) = 0$. Note that $(\bar{S}_\Psi, \bar{C}) = (\partial \cdot A^a - K^a_{\bar{C}})/\lambda$.

Observe that the action $\bar{S}_\Psi(\Phi, K)$ is no longer linear in the sources K, but contains a term that is quadratic in $K_{\bar{C}}$. This means that after integrating B out, the Φ transformations do not close off shell anymore, which the reader can verify directly. Working with the canonical formalism this problem is cured by itself, since the master equation is satisfied both before and after the integration over B.

Exercise 23 *Derive the Feynman rules of non-Abelian Yang-Mills theory, gauge-fixed as in (5.53), coupled to fermions in the fundamental representation.*

Solution. The propagators of the gauge fields and the ghosts have been given before, while those of the fermions are the same as in (1.107), multiplied by δ_{ij}. The vertices are

$$= ig f^{abc} q^{\mu},$$

$$= -g\gamma^{\mu}(T^a)_{ij},$$

$$= ig f^{abc}(\delta_{\mu\nu}(k_{1\rho} - k_{2\rho}) + \delta_{\mu\rho}(k_{3\nu} - k_{1\nu}) + \delta_{\nu\rho}(k_{2\mu} - k_{3\mu})),$$

(5.56)

$$= -g^2 \left[f^{eab} f^{ecd}(\delta_{\mu\rho}\delta_{\nu\sigma} - \delta_{\mu\sigma}\delta_{\nu\rho}) \right.$$

$$+ f^{ead} f^{ebc}(\delta_{\mu\nu}\delta_{\rho\sigma} - \delta_{\mu\rho}\delta_{\nu\sigma})$$

$$\left. + f^{eac} f^{ebd}(\delta_{\mu\nu}\delta_{\rho\sigma} - \delta_{\mu\sigma}\delta_{\nu\rho}) \right].$$

Exercise 24 *Prove that the term $\int R^{\alpha} K_{\alpha}$ can be written as (S, χ) for a local functional χ, and find χ.*

Solution. Consider the canonical transformation generated by

$$\mathcal{F}(\Phi, K') = \int \Phi^{\alpha} K'_{\alpha} + (e^{\varsigma} - 1) \int C^a K_C^{a\prime} + \left(e^{-\varsigma} - 1 \right) \int \bar{C}^a K_{\bar{C}}^{a\prime}. \quad (5.57)$$

Let $S_{\varsigma}(\Phi, K)$ denote the rescaled action. Expanding (5.44) to the first order in ς, we obtain

$$\Phi^{\alpha\prime} = \Phi^{\alpha} - \varsigma \frac{\delta\chi}{\delta K_{\alpha}} + \mathcal{O}(\varsigma^2), \qquad K'_{\alpha} = K_{\alpha} + \varsigma \frac{\delta\chi}{\delta\Phi_{\alpha}} + \mathcal{O}(\varsigma^2),$$

where

$$\chi = \int (\bar{C}^a K_{\bar{C}}^a - C^a K_C^a).$$

Expanding the transformed action, we get

$$S_{\varsigma}(\Phi, K) = S(\Phi', K') = S(\Phi, K) - \varsigma (S, \chi) + \mathcal{O}(\varsigma^2). \quad (5.58)$$

Now, the transformation rescales the ghosts by a factor e^{ς}, the antighosts by the reciprocal factor $e^{-\varsigma}$, and their sources K_C and $K_{\bar{C}}$ by $e^{-\varsigma}$ and e^{ς}, respectively. Applied to (5.31) and (5.33), even after including the gauge fixing (5.51), it is equivalent to rescaling all the sources K by e^ς, which gives $S_\varsigma(\Phi, K) = S(\Phi, e^\varsigma K)$. Differentiating this equation and (5.58) with respect to ς and setting $\varsigma = 0$ we get

$$\int R^\alpha K_\alpha = (S, \chi) .\qquad(5.59)$$

The reader is invited to check this formula explicitly in both QED and Yang-Mills theory. This result teaches us that $\int R^\alpha K_\alpha$ is *exact* in the cohomology defined by the application $X \to (S, X)$ on the local functionals X. $\Box$

From now on, we drop the subscript Ψ in $S_\Psi(\Phi, K)$. When we write $S(\Phi, K)$ we mean the gauge-fixed action (5.51).

5.5 Generating functionals

Define the generating functionals Z and W as

$$Z(J, K) = \int [\mathrm{d}\Phi] \exp\left(-S(\Phi, K) + \int \Phi^\alpha J_\alpha\right) = \exp\left(W(J, K)\right),\quad(5.60)$$

while $\Gamma(\Phi, K)$ is the Legendre transform of $W(J, K)$ with respect to the sources J, keeping the sources K fixed:

$$\Phi^\alpha = \frac{\delta_r W(\Phi, K)}{\delta J_\alpha}, \quad J_\alpha = \frac{\delta_l \Gamma(\Phi, K)}{\delta \Phi^\alpha}, \quad \Gamma(\Phi, K) = -W(J, K) + \int \Phi^\alpha J_\alpha.$$
$$(5.61)$$

When checking these formulas, pay attention to the statistics of the fields and the sources. For example, $\Phi^\alpha J_\alpha = (-1)^{\varepsilon_\alpha} J_\alpha \Phi^\alpha$ for every α, $\delta_l W / \delta J_\alpha = (-1)^{\varepsilon_\alpha} \delta_r W / \delta J_\alpha$, etc.

Observe that Γ is the generating functional of all the one-particle irreducible diagrams, including those that have ghosts, Lagrange multipliers and sources K on their external legs. We are tacitly assuming that the integral (5.60) makes sense perturbatively. This means that the action $S(\Phi, K)$ is gauge-fixed as explained earlier.

It is apparent from (5.46) that the canonical transformations cannot be implemented as changes of field variables inside the functional integral. Indeed, they mix the fields Φ, over which we are integrating, with the external sources K. While it is legitimate to make a change of field variables

$\Phi \to \Phi'(\Phi, K)$ in the functional integral, it is not legitimate to redefine the external sources as functions of the integrated fields. Thus, when we use canonical transformations, we mean that we apply them to the action $S(\Phi, K)$, and *replace* the generating functionals Z, W and Γ with the ones defined by the transformed actions.

We can treat the correlation functions

$$\langle \mathcal{O}^{I_1}(x_1) \cdots \mathcal{O}^{I_n}(x_n) \rangle \tag{5.62}$$

of composite fields $\mathcal{O}^I(\Phi)$ by coupling them to external sources L_I, and extending the generating functionals Z and W to

$$Z(J, K, L) = \int [\mathrm{d}\Phi] \exp \left(-S(\Phi, K) + \int L_I \mathcal{O}^I(\Phi) + \int \Phi^\alpha J_\alpha \right) = \mathrm{e}^{W(J,K,L)} \tag{5.63}$$

and Γ to the Lagrange transform $\Gamma(\Phi, K, L)$ of $W(J, K, L)$ with respect to Φ. We have

$$J_\alpha = \frac{\delta_l \Gamma}{\delta \Phi^\alpha}, \qquad \frac{\delta_r W}{\delta K_\alpha} = -\frac{\delta_r \Gamma}{\delta K_\alpha}, \qquad \frac{\delta_l W}{\delta L_I} = -\frac{\delta_l \Gamma}{\delta L_I}. \tag{5.64}$$

The extension is equivalent to working with the action

$$S_L(\Phi, K, L) = S(\Phi, K) - \int L_I \mathcal{O}^I(\Phi).$$

If the composite fields $\mathcal{O}^I$ are gauge invariant, the extended action S_L satisfies the master equation.

More generally, S_L satisfies the master equation any time the $\mathcal{O}^I$'s are "S closed", i.e., they satisfy $(S, \mathcal{O}^I) = 0$. Because of the nilpotency relation (5.41), trivial solutions of this equation are the "S exact" functionals, i.e., those that can be written as $\mathcal{O}^I = (S, \chi^I)$ for some local functional χ^I. Physically, these solutions are also trivial, because, roughly speaking, they are the gauge transformations of something else. For example, the Lagrange multipliers B^a satisfy both $(S, B^a) = 0$ and $B^a = (S, \bar{C}^a)$. Obviously, they are devoid of physical content, since we have introduced them just for the purpose of gauge fixing the theory. That said, we do not need to make these distinctions right now, so from now on we just assume $(S, \mathcal{O}^I) = 0$.

Recall that in this chapter we are working at the bare level: formula (5.63) encodes the bare form of the generating functionals. For the moment, the correlation functions (5.62) are gauge invariant (and gauge independent, see below), but still divergent. At the renormalized level, the exponent in the integrand can become nonpolynomial in the sources L and K, when higher-dimensional composite fields are present.

Now, consider the change of field variables

$$\Phi^{\alpha'} = \Phi^\alpha + \theta R^\alpha = \Phi^\alpha + \theta(S, \Phi^\alpha), \tag{5.65}$$

in the functional integral (5.60), where θ is a constant anticommuting parameter. If $X(\Phi)$ is a function, or functional, of the fields, its transformation reads

$$X(\Phi') = X(\Phi) + \theta(S, X). \tag{5.66}$$

We have used $\theta^2 = 0$, which ensures that the Taylor expansion in θ stops after the first order in θ.

Formula (5.66) implies that the S-exact composite fields $\mathcal{O}^I(\Phi)$ are invariant under (5.65). The action (5.68) is also invariant:

$$S(\Phi', K) = S(\Phi, K) + \frac{\theta}{2}(S, S) = S(\Phi, K). \tag{5.67}$$

In a sense that we now explain, (5.65) is equivalent to a canonical transformation generated by

$$\mathcal{F}(\Phi, K') = \int \left(\Phi^\alpha K'_\alpha + \theta R^\alpha K'_\alpha \right).$$

Indeed, formulas (5.44) give

$$\Phi^{\alpha'} = \Phi^\alpha + \theta R^\alpha, \qquad K'_\alpha = K_\alpha - \int \frac{\delta_l R^\beta}{\delta \Phi^\alpha} K_\beta \theta. \tag{5.68}$$

The K transformation that appears here does not affect the action, because S depends on K only via the combination $-\int R^\alpha(\Phi) K_\alpha$, which gets an extra contribution equal to

$$\int R^\alpha(\Phi) \frac{\delta_l R^\beta}{\delta \Phi^\alpha} K_\beta \theta = \int (S, R^\alpha) K_\alpha \theta = \int (S, (S, \Phi^\alpha)) K_\alpha \theta = 0.$$

Hence, $S(\Phi', K') = S(\Phi', K) = S(\Phi, K)$.

We know that, using the dimensional regularization, the functional integration measure is invariant under the change of field variables (5.65), by theorem 1. There actually exists a stronger argument to prove the same result, which can be applied to a more general class of regularization techniques. Thanks to (1.102) the Jacobian superdeterminant of the change of variables is

$$J = \mathrm{sdet}\frac{\delta\Phi^{\alpha\prime}(x)}{\delta\Phi^{\beta}(y)} = \mathrm{sdet}\left(\delta^{\alpha\beta}\delta(x-y) + \frac{\delta[\theta R^{\alpha}(x)]}{\delta\Phi^{\beta}(y)}\right) = 1 + \mathrm{str}\frac{\delta[\theta R^{\alpha}(x)]}{\delta\Phi^{\beta}(y)}.$$
(5.69)

We have again used $\theta^2 = 0$. In QED the matrix

$$\frac{\delta[\theta R^{\alpha}(x)]}{\delta\Phi^{\beta}(y)} = \frac{\delta[\theta(\partial_{\mu}C, 0, B, 0, -ie\bar{\psi}C, -ieC\psi)]}{\delta(A_{\nu}, C, \bar{C}, B, \bar{\psi}, \psi)}$$

has no diagonal elements except for the block

$$\frac{\delta(ie\bar{\psi}\theta C, -ie\theta C\psi)}{\delta(\bar{\psi}, \psi)} = ie\begin{pmatrix} \theta C & 0 \\ 0 & -\theta C \end{pmatrix},$$
(5.70)

whose trace vanishes. This is due to the fact that $\bar{\psi}$ and ψ have opposite charges. Using (1.100) we see that the supertrace of (5.69) vanishes, so $J = 1$.

In non-Abelian gauge theories formulas (5.32) and (5.33) give

$$\frac{\delta(\theta R^a_{\mu})}{\delta A^b_{\nu}} = g\delta_{\mu\nu}f^{abc}\theta C^c, \qquad \frac{\delta(\theta R^a_C)}{\delta C^b} = gf^{abc}\theta C^c,$$

$$\frac{\delta_l(\theta R^i_{\psi})}{\delta\psi^j} = -gT^a_{ij}\theta C^a \quad, \qquad \frac{\delta_l(\theta R^i_{\bar{\psi}})}{\delta\bar{\psi}^j} = gT^a_{ij}\theta C^a.$$

The scalar contribution is similar to the fermion one. When the representation is not the fundamental one, it is sufficient to replace T^a with the appropriate matrices $\mathcal{T}^a$. The A and C contributions to (5.69) are zero, because f^{abc} is completely antisymmetric. If the gauge group has no Abelian factors, then $\mathrm{tr}[\mathcal{T}^a] = 0$, so the traces of the ψ and φ contributions are also zero. If the gauge group has Abelian factors, the traces $\mathrm{tr}[\mathcal{T}^a]$ are proportional to the $U(1)$ charges. They cancel out by summing the contributions of both ψ and $\bar{\psi}$, or φ and $\bar{\varphi}$, as in (5.70). Finally, the contributions of $\bar{C}$ and B are obviously zero.

Now we prove that

Theorem 10 *If the action S satisfies the master equation, the generating functionals Z and W are invariant under the infinitesimal transformation*

$$\delta_\tau K_\alpha = (-1)^{\varepsilon_\alpha+1}\theta J_\alpha, \qquad \delta_\tau J_\alpha = 0, \qquad \delta_\tau L_I = 0,$$

where θ is an anticommuting parameter.

Proof. Apply the operator δ_τ to the Z functional (5.63). Using (5.19), we see that the exponent of the integrand is changed into itself plus

$$\theta \int R^\alpha(\Phi) J_\alpha. \tag{5.71}$$

Thus, we obtain the formula

$$\delta_\tau W = \theta \left\langle \int R^\alpha(\Phi) J_\alpha \right\rangle. \tag{5.72}$$

We can prove that this average vanishes by performing the change of field variables (5.65) in (5.63). Recall that the functional measure is invariant, as are the action S and the composite fields $\mathcal{O}^I$, by formulas (5.66) and (5.67). Then, (5.65) affects only $\int \Phi^\alpha J_\alpha$, by an amount equal to (5.71), and W by an amount equal to the average of (5.71), which is precisely $\delta_\tau W$. Since a change of field variables cannot modify the result of the integral, we conclude that the right-hand side of (5.72) vanishes, i.e.,

$$\delta_\tau Z(J, K, L) = 0, \qquad \delta_\tau W(J, K, L) = 0. \tag{5.73}$$

□

Using (5.64), we can write

$$\delta_\tau W = \int \delta_\tau K_\alpha \frac{\delta_l W}{\delta K_\alpha} = (-1)^{\varepsilon_\alpha}\theta \int J_\alpha \frac{\delta_l \Gamma}{\delta K_\alpha} = \theta \int \frac{\delta_r \Gamma}{\delta \Phi^\alpha} \frac{\delta_l \Gamma}{\delta K_\alpha}. \tag{5.74}$$

Furthermore, using (5.15) and (5.73), we obtain

$$\delta_\tau W = \frac{\theta}{2}(\Gamma, \Gamma) = 0, \tag{5.75}$$

which is the master equation for Γ. We have thus proved that

Theorem 11 *If S satisfies the master equation, then Γ satisfies the master equation.*

Later we will show that the Γ master equation encodes the gauge invariance of the physical correlation functions.

When the action S is not assumed to satisfy the master equation, we can prove a more general result, which tells us that the violation of the Γ master equation $(\Gamma, \Gamma) = 0$ is given by the average of (S, S). This gives a formula that is crucial for the proofs of renormalizability, and the study of anomalies and gauge independence to all orders. Due to its importance, we call it *master identity*.

Theorem 12 *The generating functional Γ satisfies the master identity*

$$(\Gamma, \Gamma) = \langle\langle (S, S) \rangle\rangle. \tag{5.76}$$

Proof. It can be proved by going through the arguments that lead to (5.75), and making the necessary modifications. Formula (5.72) is unaffected. Instead, the change of variables (5.65) does not only affect $\int \Phi^\alpha J_\alpha$, by an amount equal to (5.71), but also $-S$, by an amount equal to $-\theta(S, S)/2$, as shown by the first equality of (5.67). Since W cannot change under a change of variables, we obtain

$$\delta_\tau W = \theta \left\langle \int R^\alpha(\Phi) J_\alpha \right\rangle = \frac{\theta}{2} \langle\langle (S, S) \rangle\rangle. \tag{5.77}$$

Formula (5.74) is also unmodified, so in the end

$$\frac{\theta}{2} \langle\langle (S, S) \rangle\rangle = \delta_\tau W = \frac{\theta}{2} (\Gamma, \Gamma).$$

Alternatively, (5.77) gives

$$\langle\langle (S, S) \rangle\rangle = 2 \int \langle R^\alpha(\Phi) \rangle J_\alpha = -2 \int \left\langle \frac{\delta_r S}{\delta K_\alpha} \right\rangle \frac{\delta_l \Gamma}{\delta \Phi^\alpha}$$

$$= 2 \int \frac{\delta_r W}{\delta K_\alpha} \frac{\delta_l \Gamma}{\delta \Phi^\alpha} = -2 \int \frac{\delta_r \Gamma}{\delta K_\alpha} \frac{\delta_l \Gamma}{\delta \Phi^\alpha} = (\Gamma, \Gamma).$$

□

From now on, we go back to assuming that the action S does satisfy the master equation.

5.6   Ward identities

Consider the change of variables (5.65) in the functional integral

$$\int [d\Phi] \mathcal{Q}(\Phi) \exp\left(-S(\Phi, K) + \int L_I \mathcal{O}^I(\phi)\right), \tag{5.78}$$

where now $\mathcal{Q}$ denotes a completely arbitrary function of the fields. It can include any string of insertions of elementary and composite fields, including ghosts and Lagrange multipliers, as well as functionals. It does not even need to be local. However, for the derivation that we give below, $\mathcal{Q}$ cannot depend on the sources K. The reason is that the functional integral is only over Φ, so the change of variables cannot affect K. Note that in (5.78) we have set the sources J for the elementary fields Φ to zero. The reason is that most sources J are not gauge invariant. By means of (5.78), we can study the correlation functions (5.62).

If S satisfies the master equation, then only $\mathcal{Q}(\Phi)$ is affected by (5.65), and we easily obtain

$$\left\langle \int R^\alpha \frac{\delta_l \mathcal{Q}}{\delta \Phi^\alpha} \right\rangle_0 = \langle (S, \mathcal{Q}) \rangle_0 = 0, \tag{5.79}$$

where the subscript 0 reminds us that the sources J for the elementary fields are set to zero.

The identity (5.79) is called *Ward identity*. Its meaning is that an object of the form $(S, \mathcal{Q})$ is zero for every physical purpose, that is to say, a completely unobservable quantity. Observe that $(S, \mathcal{Q})$ is a functional of the fields only. Replacing $\mathcal{Q}$ with $\mathcal{Q}(S, \mathcal{Q})^{n-1}$ in (5.79), it follows that

$$\langle (S, \mathcal{Q})^n \rangle_0 = 0$$

for every n. Then, if we specialize $\mathcal{Q}$ to be a local functional Ψ of fermionic statistics, we also have the identity

$$\int [d\Phi] \, e^{-S_\Psi(\Phi, K) + \int L_I \mathcal{O}^I(\phi)} = \int [d\Phi] \, e^{-S(\Phi, K) + \int L_I \mathcal{O}^I(\phi)}, \tag{5.80}$$

where S_Ψ and S are related by formula (5.49), or, which is the same, the canonical transformation (5.50). The identity (5.80) tells us that we are free to add an arbitrary functional of the form (S, Ψ) to the action, and no correlation function (5.62) will depend on it. We have already seen that this freedom allows us to gauge-fix the theory, by choosing a Ψ of the form (5.48).

This proves that

Theorem 13 *The correlation functions (5.62) are invariant under the canonical transformations of the form (5.50), for an arbitrary local $\Psi(\Phi)$.*

Since the most general canonical transformation is a combination of a canonical transformation of type (5.50) and a change of variables for the fields Φ, we conclude that

Theorem 14 *The physical quantities are invariant under the most general canonical transformations.*

Among the freedom we have, we can replace $\partial_\mu A_\mu$ in (5.48) by another gauge-fixing function $\mathcal{G}(A)$. From the arbitrariness of $\Psi(\Phi)$ and theorem 13, we conclude that

Theorem 15 *The correlation functions (5.62) are gauge-independent,*

that is to say, they are independent of the gauge fixing. Even if we stick to the same $\mathcal{G}(A)$, they are independent of the gauge-fixing parameter λ that appears in (5.48).

Note that the notion of gauge independence does not coincide with the notion of gauge invariance. A gauge invariant quantity is a quantity that does not change when a gauge transformation is applied to it. A gauge independent quantity is a quantity that does not change when we modify the gauge-fixing function $\mathcal{G}(A)$ that is used to define the functional integral.

In the same way as the gauge invariance of the action S is encoded into the master equation $(S, S) = 0$, the gauge invariance of the Γ functional is encoded into the master equation $(\Gamma, \Gamma) = 0$. Indeed, we can write

$$0 = \frac{\theta}{2}(\Gamma, \Gamma) = \int \Delta\Phi^\alpha \frac{\delta_l \Gamma}{\delta\Phi^\alpha}, \qquad \Delta\Phi^\alpha \equiv -\theta\frac{\delta_r \Gamma}{\delta K_\alpha}.$$

This result tells us that Γ is invariant under the transformation $\Phi^\alpha \to \Phi^\alpha + \Delta\Phi^\alpha$, which we can interpret as the gauge symmetry of the fields Φ^α. Note that the variations $\Delta\Phi^\alpha$ are nonlocal functionals. They are equal to θR^α plus radiative corrections.

Gauge independence ensures that the values of the physical correlation functions, such as (5.62), are the same with any gauge choice. In particular, they coincide with the values we would find, for example, in the Coulomb gauge (4.9), where only the physical degrees of freedom propagate. For this reason, gauge independence proves unitarity.

We still have to prove that the theory is renormalizable. So far, we have been working with quantities that are gauge invariant and gauge independent, but still divergent. We must show that the subtraction of the divergences can be organized so as to preserve the properties proved above. The basic reason is as follows. The theory is gauge invariant at the bare level, as we have just proved. The dimensional regularization technique is manifestly gauge invariant. The divergent parts must be gauge invariant as well, because, when we expand in powers of ε and $1/\varepsilon$, every term of the expansion is separately invariant. The renormalized quantities are then invariant, because they are the bare quantities minus their divergent parts. The same line of arguments works for gauge independence.

Chapter 6

Quantum electrodynamics

In this chapter we study quantum electrodynamics, and prove its renormalizability to all orders. Since the action does not contain chiral fermions, the properties we have derived in the previous chapter, such as the master equation (5.17), hold in arbitrary complex D dimensions. In particular, the Lagrangian

$$\mathcal{L}_0 = \frac{1}{4}F_{\mu\nu}^2 + \bar{\psi}(\slashed{\partial} + ie\slashed{A} + m)\psi \tag{6.1}$$

is gauge invariant in D dimensions, and the dimensionally regularized gauge-fixed extended action

$$S(\Phi, K) = \int \mathcal{L}_{\text{tot}} + \int (K_\mu \partial_\mu C + ie\bar{\psi}CK_{\bar{\psi}} + ieK_\psi C\psi - BK_{\bar{C}}), \tag{6.2}$$

where

$$\mathcal{L}_{\text{tot}} = \mathcal{L}_0 - \frac{\lambda}{2}B^2 + B\partial \cdot A - \bar{C}\Box C, \tag{6.3}$$

satisfies $(S, S) = 0$ identically.

After integrating B out, the Feynman rules are

$$\tag{6.4}$$

where the wiggled line denotes the photon. We do not need rules for the ghosts, since they decouple.

The first thing to note is that (6.1) does not contain all the terms that are allowed by power counting. The missing ones, such as

$$\frac{1}{2}m_\gamma^2 A_\mu^2, \qquad \frac{1}{3!}A_\mu^2 \partial_\nu A_\nu, \qquad \frac{1}{4!}(A_\mu^2)^2, \tag{6.5}$$

are forbidden by gauge invariance. In principle, renormalization might generate them, at one loop or higher orders. More precisely, it might be necessary to introduce the vertices (6.5) as counterterms, to remove the divergences proportional to them. If that happened, renormalization would ruin the gauge invariance of the theory. We need to prove that, instead, the divergent parts of the Feynman diagrams are gauge invariant, and can be removed by redefining the ingredients (fields, sources and parameters) of the tree-level action $S(\Phi, K)$. Fortunately, in most cases, which include QED, renormalization and gauge invariance are compatible with each other.

For the moment, we just assume that this compatibility holds and work out some consequences. The renormalizability of (6.3) is proven in section 6.2.

Exercise 25 *Using the dimensional regularization, prove by explicit computation that the photon four-point function* $\langle A_\mu(x)A_\nu(y)A_\rho(z)A_\sigma(w)\rangle$ *is one-loop convergent.*

Solution. By power counting and locality, the divergent part is just a constant, so it can be calculated at vanishing external momenta. Although the divergent part is independent of the mass, we keep m nonzero, because the limit where the external momenta and the masses tend to zero at the same time cannot be taken inside the integral, in dimensional regularization. We have the diagram

$$(6.6)$$

plus permutations of the external legs, which means exchanges of ν, ρ and σ. The integral corresponding to (6.6) is

$$-e^4 \int \frac{\mathrm{d}^D p}{(2\pi)^D} \frac{\mathrm{tr}[(-i\not{p}+m)\gamma_\mu(-i\not{p}+m)\gamma_\nu(-i\not{p}+m)\gamma_\rho(-i\not{p}+m)\gamma_\sigma]}{(p^2+m^2)^4}.$$

The masses in the numerator can be dropped, since they contribute only to the finite part. We get

$$-e^4 \int \frac{d^D p}{(2\pi)^D} \frac{p_\alpha p_\beta p_\gamma p_\delta}{(p^2+m^2)^4} \text{tr}[\gamma_\alpha \gamma_\mu \gamma_\beta \gamma_\nu \gamma_\gamma \gamma_\rho \gamma_\delta \gamma_\sigma]. \tag{6.7}$$

By Lorentz covariance, the integral can only be proportional to $\delta_{\alpha\beta}\delta_{\gamma\delta} + \delta_{\alpha\gamma}\delta_{\beta\delta} + \delta_{\alpha\delta}\delta_{\beta\gamma}$. The factor in front of this tensor can be calculated by contracting α with β and γ with δ. We can thus write

$$\int \frac{d^D p}{(2\pi)^D} \frac{p_\alpha p_\beta p_\gamma p_\delta}{(p^2+m^2)^4} = \frac{\delta_{\alpha\beta}\delta_{\gamma\delta} + \delta_{\alpha\gamma}\delta_{\beta\delta} + \delta_{\alpha\delta}\delta_{\beta\gamma}}{D(D+2)} \int \frac{d^D p}{(2\pi)^D} \frac{(p^2)^2}{(p^2+m^2)^4}.$$

Evaluating the integral with the help of formula (A.5), then using (2.14) and (2.15) to compute the trace, we can easily find that the divergent part of (6.7) is nontrivial, and equal to

$$-\frac{8e^4}{3\varepsilon(4\pi)^2} \left(\delta_{\mu\nu}\delta_{\rho\sigma} - 2\delta_{\mu\rho}\delta_{\nu\sigma} + \delta_{\mu\sigma}\delta_{\nu\rho}\right),$$

where $\varepsilon = 4 - D$. However, the pole disappears by summing over the permutations of the external legs. Without this cancellation, there would be a divergence proportional to $(A_\mu^2)^2$, which would violate gauge invariance. This exercise is also an explicit check that the dimensional regularization is gauge invariant.

Exercise 26 *Show that the three-point function* $\langle A_\mu(x)A_\nu(y)A_\rho(z)\rangle$ *is also convergent at one loop.*

Solution. The fermion loop with three external photons has a nontrivial divergent part, which is linear in the external momenta. It is convenient to differentiate with respect to one external momentum at a time, and set the external momenta to zero afterwards (keeping the mass m in the denominators, to avoid infrared problems). The differentiated integral has a vanishing degree of divergence. Exchanging two photon legs is equivalent to flipping the arrow of the fermion loop. The divergent parts cancel out by summing the two options.

There exist more powerful methods, based on the invariance under charge conjugation, to show that, at the unregularized level, when a closed fermion loop has an odd number of photons attached to it, the two orientations of the loop compensate for each other. If we want to apply those methods in the context of the dimensional regularization, we must adapt it in a clever way, since the charge-conjugation matrix, like the matrix γ_5, does not admit a straightforward extension to D dimensions. $\square$

As usual, we have bare and renormalized versions of $\mathcal{L}_0$, which read

$$\mathcal{L}_{0\mathrm{B}} = \frac{1}{4} F_{\mu\nu\mathrm{B}}^2 + \bar{\psi}_{\mathrm{B}}(\slashed{\partial} + ie_{\mathrm{B}}\slashed{A}_{\mathrm{B}} + m_{\mathrm{B}})\psi_{\mathrm{B}} =$$

$$\mathcal{L}_{0\mathrm{R}} = \frac{1}{4} Z_A F_{\mu\nu}^2 + Z_\psi \bar{\psi}(\slashed{\partial} + ie\mu^{\varepsilon/2} Z_e Z_A^{1/2}\slashed{A} + mZ_m)\psi, \qquad (6.8)$$

having defined

$$A_{\mu\mathrm{B}} = Z_A^{1/2} A_\mu, \qquad \psi_{\mathrm{B}} = Z_\psi^{1/2}\psi, \qquad e_{\mathrm{B}} = e\mu^{\varepsilon/2} Z_e, \qquad m_{\mathrm{B}} = mZ_m. \qquad (6.9)$$

We have replaced e by $e\mu^{\varepsilon/2}$ at the tree level, to make the renormalized electric charge e dimensionless.

The renormalization of the gauge-fixing sector is rather simple. Since C and $\bar{C}$ decouple, they are not renormalized, so $C_{\mathrm{B}} = C$, $\bar{C}_{\mathrm{B}} = \bar{C}$. Moreover, B appears only quadratically in (6.3), hence no one-particle irreducible diagram with B external legs can be built. Therefore, the Lagrangian terms involving B are not renormalized either. Writing

$$B_{\mathrm{B}} = Z_B^{1/2} B, \qquad \lambda_{\mathrm{B}} = \lambda Z_\lambda, \qquad (6.10)$$

we have

$$-\frac{\lambda}{2} B^2 + B\partial \cdot A = -\frac{\lambda_{\mathrm{B}}}{2} B_{\mathrm{B}}^2 + B_{\mathrm{B}}\partial \cdot A_{\mathrm{B}} = -\frac{\lambda Z_\lambda}{2} Z_B B^2 + Z_B^{1/2} Z_A^{1/2} B\partial \cdot A, \qquad (6.11)$$

that is to say,

$$Z_B = Z_A^{-1}, \qquad Z_\lambda = Z_A. \qquad (6.12)$$

We see that B can have a nontrivial renormalization constant.

Now, let us consider the terms proportional to the sources in (6.2). The term $BK_{\bar{C}}$ is not renormalized by the argument just given. Moreover, since the ghosts decouple, no irreducible diagrams with sources K_ψ, $K_{\bar{\psi}}$ and/or K_μ on the external legs can be built. This means that the entire K sector of the solution (6.2) to the master equation is nonrenormalized, so

$$K_{\mu\mathrm{B}} = K_\mu, \qquad K_{\bar{C}\mathrm{B}} = Z_B^{-1/2} K_{\bar{C}},$$

$$K_{\bar\psi B} = Z_e^{-1} Z_\psi^{-1/2} K_{\bar\psi}, \qquad K_{\psi B} = Z_e^{-1} Z_\psi^{-1/2} K_\psi.$$

The renormalized solution of the master equation reads

$$S_R(\Phi, K) = \int (\mathcal{L}_{0R} + \mathcal{L}_{gf} + \mathcal{L}_K) = S_B(\Phi_B, K_B),$$

where

$$\mathcal{L}_{gf} = -\frac{\lambda}{2} B^2 + B \, \partial \cdot A - \bar C \square C, \tag{6.13}$$

$$\mathcal{L}_K = K_\mu \partial_\mu C + ie\mu^{\varepsilon/2} \bar\psi C K_{\bar\psi} + ie\mu^{\varepsilon/2} K_\psi C \psi - B K_{\bar C}. \tag{6.14}$$

6.1 Ward identities

The Ward identities (5.79) allow us to derive relations among the correlation functions and the renormalization constants. Before deriving the main formulas, let us mention two simple, but useful properties concerning the functional integral over the ghosts C, $\bar C$ and the Lagrange multiplier B.

Since B does not propagate, and appears quadratically in the action, integrating over B is equivalent, to some extent that we now explain, to replace it with the solution

$$B = \frac{1}{\lambda} \partial \cdot A \tag{6.15}$$

of its own field equation. Precisely, let $X(B)$ be a local functional of B (and possibly other fields). Upon making a translation, we find

$$\langle X \rangle_B \equiv \int [dB] X(B) \exp\left(\frac{\lambda}{2} \int B^2 - \int B \partial \cdot A\right)$$

$$= \int [dB] X \left(B + \frac{\partial \cdot A}{\lambda}\right) \exp\left(\frac{\lambda}{2} \int B^2 - \frac{1}{2\lambda} \int (\partial \cdot A)^2\right).$$

Now, expand $X(B)$ in powers of B. Observe that each odd power integrates to zero. On the other hand, the nonvanishing even powers give $\delta(0)$'s or derivatives of $\delta(0)$'s, which, by formulas (2.12) and (2.13), vanish using dimensional regularization. For example, normalizing $\langle 1 \rangle_B$ to one at $A_\mu = 0$, we find

$$\int [dB] B(x) \partial_\mu B(x) \exp\left(\frac{\lambda}{2} \int B^2\right) = -\frac{1}{2\lambda} \partial_\mu \delta(x - y)|_{y=x}.$$

We conclude that

$$\langle X \rangle_B = X \left(\frac{\partial \cdot A}{\lambda} \right) \exp \left(-\frac{(\partial \cdot A)^2}{2\lambda} \right). \tag{6.16}$$

Another useful property is that, since the ghosts decouple, the correlation functions involving ghost insertions factorize, i.e.

$$\langle C(x_1) \cdots C(x_m) \bar{C}(y_1) \cdots \bar{C}(y_n) \chi \rangle$$
$$= \langle C(x_1) \cdots C(x_m) \bar{C}(y_1) \cdots \bar{C}(y_n) \rangle \langle \chi \rangle, \tag{6.17}$$

where χ is any string of elementary fields other than the ghosts, e.g.

$$\chi = A_{\mu_1}(x_1) \cdots A_{\mu_n}(x_n) \bar{\psi}(y_1) \cdots \bar{\psi}(y_m) \psi(z_1) \cdots \psi(z_m).$$

Formula (6.17) can be easily proved by writing down the expressions of the averages as functional integrals.

We obtain the first Ward identity by choosing $\Psi = \bar{C}(x) \partial \cdot A_B(y)$ in formula (5.79), which gives

$$0 = \langle B_B(x) \partial \cdot A_B(y) \rangle_0 - \langle \bar{C}(x) \Box C(y) \rangle_0.$$

We recall that the subscript 0 reminds us that the sources J are set to zero. Using the bare version of (6.16), we can replace B_B with $(\partial \cdot A_B)/\lambda_B$. Next, using

$$\langle C(y) \bar{C}(x) \rangle_0 = G_{\text{free}}(y - x) \tag{6.18}$$

where $G_{\text{free}}(y - x)$ is the solution of $-\Box G_{\text{free}}(y - x) = \delta(y - x)$, we find

$$\langle \partial \cdot A_B(x) \partial \cdot A_B(y) \rangle_0 = \lambda_B \delta(x - y).$$

Switching to renormalized quantities by means of (6.9) and (6.10), this identity becomes

$$\langle \partial \cdot A(x) \partial \cdot A(y) \rangle_0 = \frac{\lambda Z_\lambda}{Z_A} \delta(x - y). \tag{6.19}$$

Note that we have not used the relations (6.12), so far. Since the left-hand side of (6.19) is convergent, by construction, the right-hand side must also be convergent, so we must have

$$\frac{Z_\lambda}{Z_A} = \text{convergent}. \tag{6.20}$$

In the minimal subtraction scheme, every Z has the form $1 +$ poles in ε, so

$$\bar{Z}_\lambda = \bar{Z}_A. \qquad (6.21)$$

The bar over the Z's is to remind us that the renormalization constants are evaluated in the MS scheme. The result (6.21) agrees with (6.12). When we derived (6.12), indeed, we implicitly used the minimal subtraction scheme, since we concentrated on the purely divergent parts. More generally, we know that we can always "subtract" (actually, "add", in this case) arbitrary finite local counterterms. If we do this in (6.11), we end up with (6.20).

As a second example, take $\Psi = \bar{C}_B(x)\bar{\psi}_B(y)\psi_B(z)$ in (5.79). This gives

$$0 = \langle B_B(x)\bar{\psi}_B(y)\psi_B(z)\rangle_0 + ie_B\langle \bar{C}(x)\bar{\psi}_B(y)C(y)\psi_B(z)\rangle_0$$
$$-ie_B\langle \bar{C}(x)\bar{\psi}_B(y)C(z)\psi_B(z)\rangle_0.$$

Using (6.18), (6.16) and (6.17), we find

$$\frac{1}{\lambda_B}\langle \partial\cdot A_B(x)\,\bar{\psi}_B(y)\psi_B(z)\rangle_0 = -ie_B\langle \bar{\psi}_B(y)\psi_B(z)\rangle_0 \left[G_{\text{free}}(x-y) - G_{\text{free}}(x-z)\right].$$

In terms of renormalized quantities, we have

$$\frac{Z_A^{1/2}}{\lambda Z_\lambda}\langle \partial\cdot A(x)\,\bar{\psi}(y)\psi(z)\rangle_0 = -ie\mu^{\varepsilon/2}Z_e\langle \bar{\psi}(y)\psi(z)\rangle_0 \left[G_{\text{free}}(x-y) - G_{\text{free}}(x-z)\right].$$

Since the correlation functions appearing in this equation are convergent, we conclude

$$\frac{Z_A^{1/2}}{Z_\lambda Z_e} = \text{convergent}. \qquad (6.22)$$

In the minimal subtraction scheme, we have

$$\bar{Z}_A = \bar{Z}_\lambda = \bar{Z}_e^{-2}. \qquad (6.23)$$

Exercise 27 *Using the dimensional regularization, compute the renormalization of QED at one loop, and check (6.23).*

Solution. We have already checked in exercises 25 and 26 that the photon four- and three-point functions are convergent. The fermion tadpole trivially vanishes. The surviving diagrams are

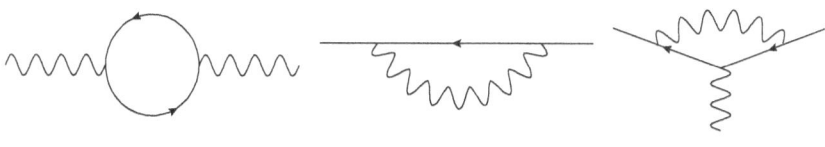

$$(6.24)$$

The first diagram is called "vacuum polarization". Its divergent part is

$$-\frac{e^2}{6\pi^2\varepsilon}(k^2\delta_{\mu\nu} - k_\mu k_\nu), \tag{6.25}$$

where k is the external momentum. From it, we find

$$Z_A = 1 - \frac{e^2}{6\pi^2\varepsilon}. \tag{6.26}$$

Note that (6.25) is transverse, namely, it vanishes when we contract it with k_μ, or k_ν. This means that the gauge-fixing term $(\partial \cdot A)^2/(2\lambda)$ is nonrenormalized, so

$$Z_\lambda = 1 - \frac{e^2}{6\pi^2\varepsilon} = Z_A,$$

in agreement with the first Ward identity (6.21).

The second diagram of (6.24) is the electron self-energy. Its divergent part is

$$\frac{-ie^2\lambda}{8\pi^2\varepsilon}\slashed{p} - \frac{me^2}{8\pi^2\varepsilon}(\lambda + 3), \tag{6.27}$$

where p is the external momentum, oriented according to the arrow. We find

$$Z_\psi = 1 - \frac{\lambda e^2}{8\pi^2\varepsilon}, \qquad Z_m = 1 - \frac{3e^2}{8\pi^2\varepsilon}. \tag{6.28}$$

Finally, by locality and power counting the divergent part of the vertex-diagram can be calculated at vanishing external momenta. Moreover, the masses can be dropped from the numerators. We then easily find

$$-\frac{i\lambda e^3}{8\pi^2\varepsilon}\gamma_\mu, \tag{6.29}$$

whence

$$Z_e = 1 + \frac{e^2}{12\pi^2\varepsilon} = Z_A^{-1/2}, \tag{6.30}$$

in agreement with the second Ward identity (6.23). Observe that only Z_ψ is gauge dependent. Later we will appreciate why. We will also be able to characterize the gauge dependence more precisely. □

Two interesting consequences of the Ward identities in the minimal subtraction scheme can be derived very easily.

i) The covariant derivative is not renormalized. Precisely,

$$D_\mu \equiv \partial_\mu + ie_{\rm B}A_{{\rm B}\mu} = \partial_\mu + ie\mu^{\varepsilon/2}\bar{Z}_e\bar{Z}_A^{1/2}A_\mu = \partial_\mu + ie\mu^{\varepsilon/2}A_\mu.$$

ii) The renormalization of the fields and the sources can be expressed in the form

$$\Phi_{\rm B}^{\alpha} = (\bar{Z}_{\Phi})^{1/2}\Phi^{\alpha}, \qquad K_{\alpha{\rm B}} = \bar{Z}_e^{-1}(\bar{Z}_{\Phi})^{-1/2}K_{\alpha}. \qquad (6.31)$$

Indeed, collecting all the pieces of information found so far, we have

$$
\begin{aligned}
&A_{\mu{\rm B}} = \bar{Z}_e^{-1}A_{\mu}, \qquad K_{\mu{\rm B}} = K_{\mu}, \qquad \psi_{\rm B} = \bar{Z}_{\psi}^{1/2}\psi, \\
&K_{\psi{\rm B}} = \bar{Z}_e^{-1}\bar{Z}_{\psi}^{-1/2}K_{\psi}, \qquad K_{C{\rm B}} = \bar{Z}_e^{-1}K_C, \qquad C_{\rm B} = C, \qquad\qquad (6.32) \\
&B_{\rm B} = \bar{Z}_e B, \qquad K_{B{\rm B}} = \bar{Z}_e^{-2}K_B, \qquad K_{\bar{C}{\rm B}} = \bar{Z}_e^{-1}K_{\bar{C}}, \qquad \bar{C}_{\rm B} = \bar{C}, \\
&e_{\rm B} = e\mu^{\varepsilon/2}\bar{Z}_e, \qquad m_{\rm B} = m\bar{Z}_m, \qquad \lambda_{\rm B} = \lambda\bar{Z}_e^{-2}.
\end{aligned}
$$

The renormalizations of K_B and K_C are completely arbitrary, since the action does not depend on them. We have chosen them to enforce (6.31).

We see that only three renormalization constants are independent. The meaning of (6.31) is that the renormalization of the fields and the sources K is a canonical transformation $(\Phi_{\rm B}, K_{\rm B}) \rightarrow (\Phi', K')$ generated by

$$
\begin{aligned}
\mathcal{F}(\Phi', K_{\rm B}) = \int \Big(&\bar{Z}_e^{-1}A'_{\mu}K_{\rm B}^{\mu} + \bar{Z}_{\psi}^{1/2}K_{\psi{\rm B}}\psi' + \bar{Z}_{\psi}^{1/2}\bar{\psi}'K_{\bar{\psi}{\rm B}} \\
&+ K_{C{\rm B}}C' + \bar{C}'K_{\bar{C}{\rm B}} + \bar{Z}_e B'K_{B{\rm B}}\Big),
\end{aligned}
$$

combined with a rescaling

$$\Phi' = \Phi, \qquad K' = \bar{Z}_e^{-1}K,$$

of the sources by $\bar{Z}_e^{-1}$. The complete renormalization is made of these two operations plus redefinitions

$$e_{\rm B} = e\mu^{\varepsilon/2}\bar{Z}_e, \qquad m_{\rm B} = m\bar{Z}_m, \qquad \lambda_{\rm B} = \lambda\bar{Z}_e^{-2},$$

of the electric charge e, the electron mass m and the gauge-fixing parameter λ. The λ redefinition is also a canonical transformation (a change of the gauge-fixing).

We can write the relation between the bare and the renormalized antiparentheses as

$$\bar{Z}_e^{-1}(X, Y)_{\rm B} = (X, Y). \qquad (6.33)$$

Details about nonminimal subtraction schemes are given in the next section.

6.2   Renormalizability of QED to all orders

Now we prove that quantum electrodynamics is renormalizable to all orders in a gauge invariant way. We work out the proof in the minimal subtraction scheme. Every other gauge invariant scheme can be reached by adding finite local, gauge invariant terms to the action. Consider the bare generating functional

$$Z_B(J_B, K_B) = \int [d\Phi_B] \exp\left(-S_B(\Phi_B, K_B) + \int \Phi_B^\alpha J_{\alpha B}\right) = e^{W_B(J_B, K_B)},$$
(6.34)

written in terms of bare fields and bare sources. The action S_B is the one of formula (5.31) once the subscript B is inserted everywhere.

We know that S_B satisfies the master equation, $(S_B, S_B)_B = 0$. Then theorem 11 ensures that the bare Γ functional Γ_B also satisfies $(\Gamma_B, \Gamma_B)_B = 0$. This identity implies

$$0 = -\int \frac{\delta_r \Gamma_B}{\delta K_{\alpha B}} \frac{\delta_l \Gamma_B}{\delta \Phi_B^\alpha} = \int \frac{\delta_r W_B}{\delta K_{\alpha B}} \frac{\delta_l \Gamma_B}{\delta \Phi_B^\alpha} = \int \langle R_B^\alpha(\Phi) \rangle \frac{\delta_l \Gamma_B}{\delta \Phi_B^\alpha}.$$
(6.35)

Now, observe that $\langle R_B^A(\Phi_B) \rangle = R_B^A(\Phi_B)$. This is obvious for $\Phi_B = A_B, C_B, \bar{C}_B$ and B_B, because their functions $R_B^A(\Phi_B)$ vanish or are linear in the fields themselves. It is less obvious for $\Phi_B = \bar{\psi}_B, \psi_B$. Yet, it is true even in those cases, because the ghosts decouple, so by (6.17) we have $\langle C_B \psi_B \rangle = \langle C_B \rangle \langle \psi_B \rangle$ and $\langle \bar{\psi}_B C_B \rangle = \langle \bar{\psi}_B \rangle \langle C_B \rangle$. We conclude that the bare functional Γ_B satisfies

$$\int R_B^\alpha(\Phi_B) \frac{\delta_l \Gamma_B}{\delta \Phi_B^\alpha} = 0.$$
(6.36)

More explicitly,

$$\int \left(\partial_\mu C_B \frac{\delta_l \Gamma_B}{\delta A_{\mu B}} - ie_B \bar{\psi}_B C_B \frac{\delta_l \Gamma_B}{\delta \bar{\psi}_B} + ie_B \frac{\delta_r \Gamma_B}{\delta \psi_B} C_B \psi_B + B_B \frac{\delta_l \Gamma_B}{\delta \bar{C}_B}\right) = 0.$$
(6.37)

Now we proceed inductively. Assume that the theory can be renormalized up to the nth loop order included by means of renormalization constants $\bar{Z}_{e,n}$, $\bar{Z}_{\psi,n}$ and $\bar{Z}_{m,n}$, and that the renormalized action reads

$$S_n(\Phi, K) = \frac{1}{4} \int \bar{Z}_{e,n}^{-2} F_{\mu\nu}^2 + \int \bar{Z}_{\psi,n} \bar{\psi}(\partial\!\!\!/ + ie\mu^{\varepsilon/2} A\!\!\!/ + m\bar{Z}_{m,n})\psi + \int (\mathcal{L}_{gf} + \mathcal{L}_K),$$

in the minimal subtraction scheme. The relations between the bare and the renormalized quantities are (6.32) with $\bar{Z}_e, \bar{Z}_\psi, \bar{Z}_m \to \bar{Z}_{e,n}, \bar{Z}_{\psi,n}, \bar{Z}_{m,n}$. Let

$\Gamma_n(\Phi, K, e, m, \lambda) = \Gamma_B(\Phi_B, K_B, e_B, m_B, \lambda_B)$ denote the n-loop renormalized generating functional of the one-particle irreducible diagrams.

We must prove that the inductive hypotheses are promoted to the $(n+1)$-th loop order. Switching formula (6.37) to the renormalized quantities, we find that all the renormalization constants simplify, apart from a common factor $\bar{Z}_e$, which we can drop. At the end, we have

$$\int \left(\partial_\mu C \frac{\delta_l \Gamma_n}{\delta A^\mu} - ie\mu^{\varepsilon/2} \bar\psi C \frac{\delta_l \Gamma_n}{\delta\bar\psi} + ie\mu^{\varepsilon/2} \frac{\delta_r \Gamma_n}{\delta\psi} C\psi + B \frac{\delta_l \Gamma_n}{\delta\bar C} \right) = 0. \quad (6.38)$$

We know that the gauge-fixing sector (6.13) and the K sector (6.14) do not renormalize. Actually they do not receive any radiative corrections, because no diagrams can be constructed with those sets of external legs. Thus, we have

$$\Gamma_n(\Phi, K) = \tilde\Gamma_n(A, \bar\psi, \psi) + \int (\mathcal{L}_{\text{gf}} + \mathcal{L}_K).$$

Inserting this formula in (6.38), we get

$$\int \left(\partial_\mu C \frac{\delta_l \tilde\Gamma_n}{\delta A^\mu} - ie\mu^{\varepsilon/2} \bar\psi C \frac{\delta_l \tilde\Gamma_n}{\delta\bar\psi} + ie\mu^{\varepsilon/2} \frac{\delta_r \tilde\Gamma_n}{\delta\psi} C\psi \right) = 0.$$

Multiplying by a constant anticommuting parameter θ to the left and identifying θC with an ordinary (commuting) function Λ, we obtain that $\tilde\Gamma_n$ is gauge invariant, that is to say

$$\delta_\Lambda \tilde\Gamma_n = 0, \quad (6.39)$$

where δ_Λ is given by (4.7) with $e \to e\mu^{\varepsilon/2}$.

To keep track of the orders of the loop expansion, we reintroduce $\hbar$ for a moment. Define

$$\tilde\Gamma_n = \sum_{k=0}^{\infty} \hbar^k \tilde\Gamma_n^{(k)}. \quad (6.40)$$

Observe that δ_Λ is independent of $\hbar$. Taking the $(n+1)$-th order of (6.39), we obtain

$$\delta_\Lambda \tilde\Gamma_n^{(n+1)} = 0. \quad (6.41)$$

By the inductive assumption, Γ_n and $\tilde\Gamma_n$ are convergent up to the n-th order included. We express this fact by writing $\tilde\Gamma_n^{(k)} < \infty$ for $k \leqslant n$. Instead, $\tilde\Gamma_n^{(n+1)}(A, \bar\psi, \psi)$ is the sum of a divergent part, which we denote by

$\tilde{\Gamma}_{n\,\mathrm{div}}^{(n+1)}(A, \bar{\psi}, \psi)$, and a finite part. By the convergence of the functionals $\tilde{\Gamma}_n^{(k)}$ with $k \leqslant n$, all the Feynman diagrams of order $\hbar^{n+1}$ that contribute to $\tilde{\Gamma}_n$ are accompanied by appropriate counterterms, which subtract their subdivergences. Then, the theorem about the locality of the counterterms ensures that $\tilde{\Gamma}_{n\,\mathrm{div}}^{(n+1)}$ is a local functional. Taking the divergent part of (6.41) (i.e. its poles in ε), we obtain that $\tilde{\Gamma}_{n\,\mathrm{div}}^{(n+1)}$ is gauge invariant:

$$\delta_\Lambda \tilde{\Gamma}_{n\,\mathrm{div}}^{(n+1)} = 0. \tag{6.42}$$

Ultimately, $\tilde{\Gamma}_{n\,\mathrm{div}}^{(n+1)}(A, \bar{\psi}, \psi)$ is a local, gauge invariant functional, equal to the integral of some local function $\Upsilon_{n+1}(A, \bar{\psi}, \psi)$ of dimension D, which is determined up to total derivatives, and whose gauge variation $\delta_\Lambda \Upsilon_{n+1}$ is at most a total derivative.

Now we use power counting. With the Lorenz gauge-fixing, the photon propagator behaves correctly for large momenta. Moreover, the theory does not contain parameters of negative dimensions. These facts ensure that the function $\Upsilon_{n+1}(A, \psi, \bar{\psi})$ is a linear combination of the local terms of dimensions $\leqslant D$ that can be built with the fields A, ψ and $\bar{\psi}$ and their derivatives. Such terms are F^2, $(\partial \cdot A)^2$, $\bar{\psi} \partial\!\!\!/ \psi$, $\bar{\psi} A\!\!\!/ \psi$ and $\bar{\psi}\psi$, plus (6.5), up to total derivatives. We cannot use the tensor $\varepsilon^{\mu\nu\rho\sigma}$, nor the matrix γ_5, since the Feynman rules do not contain them. Finally, (6.42) reduces the list to the gauge-invariant combinations F^2, $\bar{\psi} D\!\!\!\!/ \psi$ and $\bar{\psi}\psi$, so we can write

$$\Upsilon_{n+1} = a_{n+1} F_{\mu\nu}^2 + b_{n+1}\bar{\psi} D\!\!\!\!/ \psi + c_{n+1} m\bar{\psi}\psi, \qquad \tilde{\Gamma}_{n\,\mathrm{div}}^{(n+1)} = \int \Upsilon_{n+1}, \tag{6.43}$$

for suitable divergent coefficients a_{n+1}, b_{n+1} and c_{n+1}.

These divergences can be subtracted by means of new renormalization constants

$$\bar{Z}_{e,n+1} = (\bar{Z}_{e,n}^{-2} - a_{n+1})^{-1/2}, \qquad \bar{Z}_{\psi,n+1} = \bar{Z}_{\psi,n} - b_{n+1},$$
$$\bar{Z}_{m,n+1} = (\bar{Z}_{\psi,n} - b_{n+1})^{-1}(\bar{Z}_{\psi,n}\bar{Z}_{m,n} - c_{n+1}).$$

That is to say, the renormalized action

$$S_{n+1}(\Phi, K) = \frac{1}{4}\int \bar{Z}_{e,n+1}^{-2} F_{\mu\nu}^2 + \int \bar{Z}_{\psi,n+1}\bar{\psi}(\partial\!\!\!/ + ie\mu^{\varepsilon/2} A\!\!\!/ + m\bar{Z}_{m,n+1})\psi$$
$$+ \int (\mathcal{L}_{\mathrm{gf}} + \mathcal{L}_K) = S_n(\Phi, K) - \tilde{\Gamma}_{n\,\mathrm{div}}^{(n+1)}$$

produces a generating functional Γ_{n+1} that is convergent up to and including $n+1$ loops. Indeed, since the actions $S_{n+1}(\Phi, K)$ and $S_n(\Phi, K)$ differ by $\mathcal{O}(\hbar^{n+1})$, the Feynman diagrams of order n or less are exactly the same, which ensures $\Gamma_{n+1} = \Gamma_n + \mathcal{O}(\hbar^{n+1})$. Moreover, at order $n+1$ we have exactly the same diagrams, plus the vertices of $-\tilde{\Gamma}_{n\mathrm{div}}^{(n+1)}$, which subtract the overall divergent parts. In conclusion,

$$\Gamma_{n+1} \equiv \sum_{k=0}^{\infty} \hbar^k \tilde{\Gamma}_{n+1}^{(k)} = \Gamma_n - \tilde{\Gamma}_{n\mathrm{div}}^{(n+1)} + \mathcal{O}(\hbar^{n+2}),$$

that is to say, $\tilde{\Gamma}_{n+1}^{(k)} = \tilde{\Gamma}_n^{(k)} < \infty$ for $k \leqslant n$ and $\tilde{\Gamma}_{n+1}^{(n+1)} = \tilde{\Gamma}_n^{(n+1)} - \tilde{\Gamma}_{n\mathrm{div}}^{(n+1)} < \infty$.

This result extends the inductive hypotheses to $n+1$ loops, as we wanted. Iterating the argument to $n = \infty$, the map relating the bare and the renormalized quantities is (6.32) with $\bar{Z}_e = \bar{Z}_{e,\infty}$, $\bar{Z}_\psi = \bar{Z}_{\psi,\infty}$ and $\bar{Z}_m = \bar{Z}_{m,\infty}$. The renormalized action is

$$S_R = S_\infty = \frac{1}{4} \int \bar{Z}_e^{-2} F_{\mu\nu}^2 + \int \bar{Z}_\psi \bar{\psi} (\slashed{\partial} + ie\mu^{\varepsilon/2}\slashed{A} + m\bar{Z}_m)\psi + \int (\mathcal{L}_{\mathrm{gf}} + \mathcal{L}_K) \tag{6.44}$$

and the renormalized generating functional of the one-particle irreducible correlation functions is

$$\Gamma_R(\Phi, K) = \Gamma_\infty(\Phi, K) = \tilde{\Gamma}_\infty(A, \bar{\psi}, \psi) + \int (\mathcal{L}_{\mathrm{gf}} + \mathcal{L}_K). \tag{6.45}$$

Moreover, we have
(i) $\bar{Z}_e^{-1}(X,Y)_\mathrm{B} = (X,Y)$;
(ii) $(S_R, S_R) = 0$;
(iii) $(\Gamma_R, \Gamma_R) = 0$.

Point (i) follows from (6.32), as shown in (6.33). Point (ii) follows from $(S_\mathrm{B}, S_\mathrm{B})_\mathrm{B} = 0$ and point (i). It can also be verified immediately by using (6.44). Point (iii) follows from point (ii) and theorem 11.

So far, we have worked in the absence of composite fields, which is enough to study the S matrix. When we include gauge-invariant composite fields $\mathcal{O}^I(\phi)$, built with the elementary physical fields $\phi \equiv (A, \psi, \bar{\psi})$, both the gauge-fixing sector and the K sector continue to stay uncorrected, because no nontrivial diagrams affecting them can be built. The derivation given above is unmodified up to and including (6.42).

Let $\mathcal{O}^I(\phi, e\mu^{\varepsilon/2})$ denote a basis of gauge-invariant composite fields, which includes the identity. The $\mathcal{O}^I$'s may depend on e by gauge invariance, but there is no need to assume that they depend on m. The bare action is extended to

$$S_{\mathrm{B}}(\Phi_{\mathrm{B}}, K_{\mathrm{B}}, L_{\mathrm{B}}) = S_{\mathrm{B}}(\Phi_{\mathrm{B}}, K_{\mathrm{B}}) - \int L_{\mathrm{B}}^I \mathcal{O}^I(\phi_{\mathrm{B}}, e_{\mathrm{B}}).$$

Clearly, the master equation $(S_{\mathrm{B}}, S_{\mathrm{B}})_{\mathrm{B}} = 0$ is still satisfied. We write the n-loop renormalized action as

$$S_n(\Phi, K, L) = S_n(\Phi, K) - \int f_n^I(L) \mathcal{O}^I(\phi, e\mu^{\varepsilon/2}), \qquad (6.46)$$

where $f_n^I(L)$ are local functions, to be determined, of the form L^I + poles in ε, with $f_0^I(L) = L^I$. Obviously, $(S_n, S_n) = 0$.

The sources L that multiply the composite fields of dimensions > 4 have negative dimensions in units of mass. This means that the divergent part $\tilde{\Gamma}_{n\,\mathrm{div}}^{(n+1)}$ is no longer restricted by power counting. Nevertheless, we can write

$$\tilde{\Gamma}_{n\,\mathrm{div}}^{(n+1)}(\phi, L) = \int \Upsilon_{n+1} + \int h_{n+1}^I(L) \mathcal{O}^I(\phi, e\mu^{\varepsilon/2}),$$

where the divergent functions $h_{n+1}^I(L) = \mathcal{O}(L)$ are local. As before, the divergent terms of Υ_{n+1} can be reabsorbed into the constants $\bar{Z}_e$, $\bar{Z}_\psi$ and $\bar{Z}_m$. Instead, the L-dependent divergent part can be reabsorbed by defining

$$f_{n+1}^I(L) = f_n^I(L) + h_{n+1}^I(L). \qquad (6.47)$$

The $(n+1)$-renormalized action $S_{n+1}(\Phi, K, L)$ has the form (6.46) with $n \to n+1$. Indeed, we have

$$S_{n+1}(\Phi, K, L) = S_n(\Phi, K, L) - \tilde{\Gamma}_{n\,\mathrm{div}}^{(n+1)}(\phi, L),$$

which ensures that the Γ functional $\Gamma_{n+1}(\Phi, K, L)$ is renormalized up to and including $n+1$ loops.

Iterating the argument to $n = \infty$, we find the renormalized action $S_R(\Phi, K, L) = S_\infty(\Phi, K, L)$ and the renormalized Γ functional $\Gamma_R(\Phi, K, L) = \Gamma_\infty(\Phi, K, L)$, which satisfy the properties (i), (ii) and (iii) listed above.

The solutions of (6.47), and the relations between the bare sources L_{B}^I and the renormalized sources L^I, are

$$f_\infty^I(L) = L^I + \sum_{j=1}^{\infty} h_j^I(L), \qquad L_{\mathrm{B}}^I \equiv f_\infty^J(L)(d^{-1})^{JI},$$

where the constant matrices $d^{IJ} = \delta^{IJ}$ + poles in ε are defined by

$$\mathcal{O}^I(\phi_B, e_B) = \mathcal{O}^I(\bar{Z}_\phi^{1/2}\phi, e\mu^{\varepsilon/2}\bar{Z}_e) \equiv d^{IJ}\mathcal{O}^I(\phi, e\mu^{\varepsilon/2}).$$

The renormalized fields $[\mathcal{O}^I]$ are minus the derivatives of $S_R(\Phi, K, L)$ with respect to L^I, calculated at $L = 0$. The insertions of renormalized fields into the correlation functions (5.62) are studied by differentiating the generating functional $W_R(J, K, L)$ with respect to the appropriate sources L^I, and then setting the sources L to zero.

Chapter 7

Non-Abelian gauge field theories

In this chapter we use the Batalin-Vilkovisky formalism to prove the renormalizability of Yang-Mills theory to all orders in the perturbative expansion. We concentrate on gauge theories with a simple gauge group, since the generalization to product groups is straightforward. We also assume that the theories are parity invariant, which ensures that the classical Lagrangian does not contain the matrix γ_5 and the tensor $\varepsilon^{\mu\nu\rho\sigma}$.

As usual, we denote the bare fields and the bare sources with Φ_B and K_B, respectively. We write the bare action and the bare Γ functional, defined according to (5.63), as $S_B(\Phi_B, K_B, L_B, \zeta_B, \xi_B)$ and $\Gamma_B(\Phi_B, K_B, L_B, \zeta_B, \xi_B)$, where ζ denotes the physical parameters, ξ are the gauge-fixing parameters and L are the sources for the gauge-invariant composite fields. At $L = 0$ the bare action can be read from (5.34) and (5.33), or (5.38) and (5.37), plus the gauge-fixing terms of (5.52), plus possibly (4.54) and (4.55), all the quantities appearing in those formulas being interpreted as bare. After a partial integration in the ghost sector, and omitting the subscripts $_B$, we write the action for definiteness as

$$
S(\Phi, K) = \frac{1}{4} \int F_{\mu\nu}^{a\,2} - \frac{\lambda}{2} \int (B^a)^2 + \int B^a \partial \cdot A^a - \int B^a K_{\bar{C}}^a
$$
$$
+ \int (K_\mu^a + \partial_\mu \bar{C}^a)(\partial_\mu C^a + g f^{abc} A_\mu^b C^c) + \frac{g}{2} \int f^{abc} C^b C^c K_C^a \qquad (7.1)
$$
$$
+ \int \bar{\psi}^I (\delta_{IJ} \slashed{\partial} + g T_{IJ}^a \slashed{A} + m \delta_{IJ}) \psi^J + g \int \left(\bar{\psi}^I T_{IJ}^a C^a K_\psi^J + K_{\bar{\psi}}^I T_{IJ}^a C^a \psi^J \right),
$$

225

From (5.17) we have the master equation

$$(S_{\mathrm{B}}, S_{\mathrm{B}})_{\mathrm{B}} = 0, \tag{7.2}$$

which implies, according to theorem 11,

$$(\Gamma_{\mathrm{B}}, \Gamma_{\mathrm{B}})_{\mathrm{B}} = 0. \tag{7.3}$$

As usual, renormalizability is proved by proceeding inductively. The simplest proof amounts to subtracting the counterterms "as they come" in the minimal subtraction scheme. We will see in a moment what this means. We do not need to preserve the master equation at each step of the subtraction. It is more practical to allow for certain violations, as long as they are of higher orders.

Let S_n and Γ_n denote the action and the Γ functional renormalized up to and including n loops. Expand Γ_n as

$$\Gamma_n = \sum_{k=0}^{\infty} \hbar^k \Gamma_n^{(k)}, \tag{7.4}$$

where $\Gamma_n^{(k)}$ collects the k-loop contributions. Assume the inductive hypotheses

$$S_n = S_0 + \text{poles}, \qquad (S_n, S_n) = \mathcal{O}(\hbar^{n+1}), \qquad \Gamma_n^{(k)} < \infty \quad \forall k \leqslant n. \tag{7.5}$$

The last requirement is just the statement that Γ_n is convergent up to and including n loops. Clearly, the inductive assumptions are trivially satisfied at $n = 0$. In particular, S_0 coincides with the bare action S_{B}, upon dropping the subscripts $_{\mathrm{B}}$ and attaching a factor $\mu^{\varepsilon/2}$ to the coupling g. Thus, formula (7.2) ensures $(S_0, S_0) = 0$. Using the master identity (5.76), we have

$$(\Gamma_n, \Gamma_n) = \langle\!\langle (S_n, S_n) \rangle\!\rangle. \tag{7.6}$$

By the second assumption of (7.5), the antiparenthesis (S_n, S_n) is a local functional that starts from order $\hbar^{n+1}$. Then, the contributions of order $\hbar^{n+1}$ to $\langle\!\langle (S_n, S_n) \rangle\!\rangle$ coincide with the contributions of order $\hbar^{n+1}$ to (S_n, S_n). We denote them by $(S_n, S_n)|_{n+1}$. By $(S_0, S_0) = 0$ and the first assumption of (7.5), they are made of poles in ε.

Now we use the expansion (7.4) and think of (7.6) diagrammatically, as shown in (5.43). The order $\hbar^{n+1}$ of (7.6) gives

$$\sum_{k=0}^{n+1} \left(\Gamma_n^{(k)}, \Gamma_n^{(n+1-k)} \right) = (S_n, S_n)|_{n+1} . \tag{7.7}$$

We know that the functionals $\Gamma_n^{(k)}$ are convergent for $k \leqslant n$, by the inductive assumption. Taking the divergent part of (7.7), we obtain

$$2 \left(\Gamma_n^{(0)}, \Gamma_{n \text{ div}}^{(n+1)} \right) = (S_n, S_n)|_{n+1} , \tag{7.8}$$

where $\Gamma_{n \text{ div}}^{(n+1)}$ is the order-$\hbar^{n+1}$ divergent part of Γ_n. By the third inductive assumption (7.5), all the subdivergences are subtracted away, so $\Gamma_{n \text{ div}}^{(n+1)}$ is a local functional. Now, $\Gamma_n^{(0)}$ coincides with the classical action S_0, so (7.8) becomes

$$(S_0, \Gamma_{n \text{ div}}^{(n+1)}) = \frac{1}{2} (S_n, S_n)|_{n+1} . \tag{7.9}$$

At this point, define

$$S_{n+1} = S_n - \Gamma_{n \text{ div}}^{(n+1)} . \tag{7.10}$$

The first inductive assumption of the list (7.5) is clearly promoted to S_{n+1}. Formulas (7.10) and (7.9) give

$$(S_{n+1}, S_{n+1}) = (S_n, S_n) - 2 \left(S_n, \Gamma_{n \text{ div}}^{(n+1)} \right) + \left(\Gamma_{n \text{ div}}^{(n+1)}, \Gamma_{n \text{ div}}^{(n+1)} \right) = \mathcal{O}(\hbar^{n+2}),$$

so the second assumption of (7.5) is also promoted to S_{n+1}. Finally, the (irreducible) diagrams built with the vertices of S_{n+1} coincide with the diagrams of S_n, plus new diagrams containing the vertices of $-\Gamma_{n \text{ div}}^{(n+1)}$. The new diagrams are at least of order $\hbar^{n+1}$, so

$$\Gamma_n^{(k)} = \Gamma_{n+1}^{(k)} \qquad \forall k \leqslant n.$$

Moreover, the new diagrams of order $\hbar^{n+1}$ are given exactly by $-\Gamma_{n \text{ div}}^{(n+1)}$: a vertex of $-\Gamma_{n \text{ div}}^{(n+1)}$ contributes only by itself, since it is already of order $\hbar^{n+1}$, and every loop raises the order by one unit. Thus,

$$\Gamma_{n+1}^{(n+1)} = \Gamma_n^{(n+1)} - \Gamma_{n \text{ div}}^{(n+1)} < \infty,$$

which promotes the third inductive assumption of (7.5) to Γ_{n+1}.

At this point, formulas (7.5) and (7.6) imply that the renormalized action $S_R \equiv S_\infty$ and the renormalized generating functional $\Gamma_R \equiv \Gamma_\infty$ satisfy

$$(S_R, S_R) = 0, \qquad (\Gamma_R, \Gamma_R) = 0. \tag{7.11}$$

Now we study the renormalized action S_R in detail.

Proposition 16 *The renormalized action is independent of K_B^a and depends on B^a, $K_{\bar{C}}^a$ only by means of the terms*

$$\int \left(-\frac{\lambda}{2}(B^a)^2 + B^a \partial \cdot A^a - B^a K_{\bar{C}}^a \right),$$

which are nonrenormalized.

Proof. The classical action (7.1) satisfies these properties. Then, no one-particle irreducible diagrams with external legs B^a, K_B^a and $K_{\bar{C}}^a$ can be built, so no counterterms can depend on B^a, K_B^a or $K_{\bar{C}}^a$. Note, in particular, that the absence of vertices with B^a legs is due to the linearity of the gauge fixing $\mathcal{G}^a$ in A_μ^a.

Proposition 17 *The renormalized action depends on $\bar{C}$ and K_a^μ only via the combination*

$$K_a^\mu + \partial_\mu \bar{C}^a. \tag{7.12}$$

Proof. The classical action (7.1) satisfies the property. The vertex with an antighost leg $\bar{C}^a$ has a derivative ∂_μ acting on it, so it actually contains $\partial_\mu \bar{C}^a$. An almost identical vertex is the one where $\partial_\mu \bar{C}^a$ is replaced by K_a^μ. Given a diagram G with a K_a^μ external leg, there exists an almost identical diagram G', which differs from G just because the leg K_a^μ is replaced by a leg $\partial_\mu \bar{C}^a$. The converse is also true. A diagram with an arbitrary number of K_a^μ and $\partial_\mu \bar{C}^a$ external legs has almost identical partners, where the K_a^μ and $\partial_\mu \bar{C}^a$ external legs are converted in all possible ways. Then the renormalized action satisfies the property stated in the proposition.

Proposition 18 *At $L = 0$, the renormalized action is linear in K.*

Proof. From proposition 16 we know that the dependence on K_B^a and K_C^a satisfies the property. From (5.35) and (5.36) it follows that, in the absence of composite fields, any local term that is quadratic in the remaining sources K has either dimension greater than four, or ghost number different from zero. □

The propositions just stated are compatible with the subtraction algorithm given above. In particular, formula (7.10) ensures that they hold at every step of the subtraction procedure.

Propositions 16 and 17 also hold in the presence of composite fields, because their proofs do not require arguments based on power counting. Instead, 18 does not hold at $L \neq 0$, in general, because the sources L for the composite fields can have arbitrarily large negative dimensions, as well as vanishing ghost number. Then, local Lagrangian terms with arbitrarily large powers of K can be constructed, provided we adjust their dimensions by means of powers of L, and their ghost numbers by means of powers of C. We know that, in renormalization, when a term cannot be excluded *a priori* by advocating power counting, symmetries or other general properties, it is typically generated as the divergent part of some diagram. For this reason, proposition 18 is generically false at $L \neq 0$.

For a while we argue at $L = 0$, then we generalize our arguments to $L \neq 0$. Since the renormalized action S_R is linear in the sources K at $L = 0$, we can write

$$S_R(\Phi, K) = S_R'(\Phi) - \int R_\infty^\alpha(\Phi) K_\alpha.$$

The functions $R_\infty^\alpha(\Phi)$ that multiply the sources inside S_R are the renormalized symmetry transformations of the fields. By proposition 16, ghost number conservation, locality and power counting, we must have, in the notation of formulas (5.37) and (5.39),

$$S_R(\Phi, K) = S_R'(\Phi) - \int (a\partial_\mu C_j^i + bA_{\mu k}^i C_j^k - cA_{\mu j}^k C_k^i)K_{\mu i}^j + \int hC_k^i C_j^k K_{\bar C i}^j$$
$$- \int B_j^i K_{\bar C i}^j + \int \left[\left(a_1 \psi_{k_1 j_2 \cdots j_m}^{i_1 \cdots i_n} C_{j_1}^{k_1} + \ldots + a_n \psi_{j_1 \cdots j_{m-1} k_m}^{i_1 \cdots i_n} C_{j_m}^{k_m} \right) K_{i_1 \cdots i_n}^{j_1 \cdots j_m} + \ldots \right],$$

where a, b, c, h and a_k are numerical constants. Note that, out of the three tensors of (4.42), we can only use δ_j^i. The ε tensors cannot appear, just because they are not present in the Feynman rules. It is easy to check, from formula (5.20) (the "transformations of the transformations"), that the terms proportional to $K^j_{\mu i}$ of the master equation $(S_R, S_R) = 0$ give $b = c = h$. Moreover, the terms proportional to $K^{j_1 \cdots j_m}_{i_1 \cdots i_n}$ give $a_k = h$ for every k. Writing $h = iga'$, we have

$$S_R(\Phi, K) = S'_R(\Phi) + ga' \int \left(\bar{\psi}^I T^a_{IJ} C^a K^J_\psi + K^I_\psi T^a_{IJ} C^a \psi^J \right)$$
$$- \int \left[\left(a\partial_\mu C^a + ga' f^{abc} A^b_\mu C^c \right) K^a_\mu - \frac{ga'}{2} f^{abc} C^b C^c K^a_C + B^a K^a_{\bar{C}} \right],$$

Propositions 16 and 17 ensure that the B-dependent terms are nonrenormalized, and that the terms containing the antighosts $\bar{C}^a$ can be obtained from those proportional to K^a_μ upon replacing K^a_μ with $\partial_\mu \bar{C}^a$. Then $S'_R(\Phi)$ has the form

$$S'_R(\Phi) = S_{cR}(\phi) - \frac{\lambda}{2} \int (B^a)^2 + \int B^a \partial \cdot A^a$$
$$- \int \bar{C}^a \partial_\mu \left(a\partial_\mu C^a + ga' f^{abc} A^b_\mu C^c \right) = S_{cR}(\phi) + (S_R, \Psi),$$

where the gauge fermion Ψ of formula (5.48) is nonrenormalized.

Let us focus, for simplicity, on the pure gauge theory. Writing the most general local form of $S_{cR}(A)$, it is easy to check, by explicit computation, that the most general solution of $(S_R, S_{cR}) = 0$ is

$$S_{cR}(A) = \frac{a''}{4} \int d^D x \left(a\partial_\mu A^a_\nu - a\partial_\nu A^a_\mu + ga' f^{abc} A^b_\mu A^c_\nu \right)^2,$$

where a'' is another constant. Writing

$$a = Z_C, \qquad a'' = Z_A Z_C^{-2}, \qquad a' = \mu^{\varepsilon/2} Z_g Z_A^{1/2} Z_C,$$

we finally obtain

$$S_R(\Phi, K, g, \lambda) = S_B(\Phi_B, K_B, g_B, \lambda_B), \tag{7.13}$$

with

$$\begin{aligned}
A^a_{\mu B} &= Z_A^{1/2} A^a_\mu, & C^a_B &= Z_C^{1/2} C^a, & g_B &= g\mu^{\varepsilon/2} Z_g, \\
B^a_B &= Z_A^{-1/2} B^a, & \bar{C}^a_B &= Z_C^{1/2} \bar{C}^a, & \lambda_B &= \lambda Z_A, \\
K^\mu_{aB} &= Z_C^{1/2} K^\mu_a, & K^a_{\bar{C}B} &= Z_A^{1/2} K^a_{\bar{C}}, & K^a_{\bar{C}B} &= Z_A^{1/2} K^a_{\bar{C}}.
\end{aligned} \tag{7.14}$$

The inclusion of matter is straightforward: only $S_{cR}(\phi)$ changes, since it must include all the terms of dimensions $\leqslant 4$ that are invariant with respect to the renormalized gauge transformations.

At $L \neq 0$ the renormalized action has a more involved structure, since the presence of higher-dimensional composite fields can make it nonpolynomial in Φ, K and L. In the sector $L \neq 0$, we may just subtract the counterterms as they come, according to formula (7.10). We do not need to worry about rewriting the subtraction as a redefinition of the fields, the sources and the parameters (which would require, among the other things, nonpolynomial field redefinitions of the fields Φ^α and the sources K_α, instead of simple renormalizations).

Having expressed the whole of renormalization as a redefinition of the fields, the sources and the parameters in the $L = 0$ sector, we have reached our goal, since we have showed that the renormalization program can be carried out to the very end by keeping the number of independent physical parameters finite. This is a necessary requirement to ensure that predictivity is retained. At the same time, we have been able to preserve gauge invariance and gauge independence, which are the requirements for having unitarity.

The composite fields, on the other hand, do not add physical parameters to the theory, since the sources L are just tools to simplify the derivations of various properties. Thus, we do not lose much, if we renormalize the divergences belonging to the L-dependent sector by subtracting them away just as they come, as said. When we do so, the renormalized extended action $S_R(\Phi, K, L)$ and the renormalized extended Γ functional $\Gamma_R(\Phi, K, L)$ satisfy the respective master equations, by construction.

Appendix A

Notation and useful formulas

The γ matrices in four dimensions read

$$\gamma^\mu = \begin{pmatrix} 0 & \sigma^\mu \\ \bar\sigma^\mu & 0 \end{pmatrix},$$

where $\sigma^\mu = (1, \boldsymbol{\sigma})$ and $\bar\sigma^\mu = (1, -\boldsymbol{\sigma})$ in Minkowski spacetime, $\sigma^\mu = (1, i\boldsymbol{\sigma})$ and $\bar\sigma^\mu = (1, -i\boldsymbol{\sigma})$ in Euclidean space. Moreover, $\boldsymbol{\sigma} = (\sigma^1, \sigma^2, \sigma^3)$, where

$$\sigma^1 = \begin{pmatrix} 0 & 1 \\ 1 & 0 \end{pmatrix}, \qquad \sigma^2 = \begin{pmatrix} 0 & -i \\ i & 0 \end{pmatrix}, \qquad \sigma^3 = \begin{pmatrix} 1 & 0 \\ 0 & -1 \end{pmatrix},$$

are the Pauli matrices.

In Minkowski spacetime the Fourier transform is defined as

$$\varphi(x) = \int \frac{\mathrm{d}^D p}{(2\pi)^D} e^{-ip\cdot x} \tilde\varphi(p), \qquad (A.1)$$

while in Euclidean space it is

$$\varphi(x) = \int \frac{\mathrm{d}^D p}{(2\pi)^D} e^{ip\cdot x} \tilde\varphi(p).$$

To manipulate the denominators of the Feynman diagrams, it is useful to introduce the so-called Feynman parameters, by means of the formula

$$\prod_i \frac{1}{A^{\alpha_i}} = \frac{\Gamma(\sum_i \alpha_i)}{\prod_j \Gamma(\alpha_j)} \int_0^1 \prod_i \left(\mathrm{d}x_i \, x_i^{\alpha_i - 1} \right) \frac{\delta(1 - \sum_k x_k)}{(\sum_m x_m A_m)^{\sum_n \alpha_n}}.$$

Particular cases are

$$\frac{1}{AB} = \int_0^1 dx \frac{1}{[Ax + B(1-x)]^2}, \tag{A.2}$$

$$\frac{1}{A^\alpha B^\beta} = \frac{\Gamma(\alpha+\beta)}{\Gamma(\alpha)\Gamma(\beta)} \int_0^1 dx \frac{x^{\alpha-1}(1-x)^{\beta-1}}{[Ax + B(1-x)]^{\alpha+\beta}},$$

$$\frac{1}{ABC} = 2 \int_0^1 dx \int_0^{1-x} dy \frac{1}{[Ax + By + C(1-x-y)]^3}.$$

The integration over Feynman parameters often reduces to the integral

$$\int_0^1 dx \, x^{\alpha-1}(1-x)^{\beta-1} = \frac{\Gamma(\alpha)\Gamma(\beta)}{\Gamma(\alpha+\beta)}. \tag{A.3}$$

The most frequent D-dimensional integral in Euclidean space is

$$\int \frac{d^D p}{(2\pi)^D} \frac{1}{(p^2+m^2)^\alpha} = \frac{\Gamma\left(\alpha-\frac{D}{2}\right)}{(4\pi)^{D/2}\Gamma(\alpha)}(m^2)^{\frac{D}{2}-\alpha}. \tag{A.4}$$

More generally,

$$\int \frac{d^D p}{(2\pi)^D} \frac{(p^2)^\beta}{(p^2+m^2)^\alpha} = \frac{\Gamma\left(\beta+\frac{D}{2}\right)\Gamma\left(\alpha-\beta-\frac{D}{2}\right)}{(4\pi)^{D/2}\Gamma(\alpha)\Gamma\left(\frac{D}{2}\right)}(m^2)^{\frac{D}{2}-\alpha+\beta}. \tag{A.5}$$

In Minkowsky spacetime this result reads

$$\int \frac{d^D p}{(2\pi)^D} \frac{(p^2)^\beta}{(p^2-\tilde{m}^2)^\alpha} = \frac{i(-1)^{\alpha+\beta}\Gamma\left(\beta+\frac{D}{2}\right)\Gamma\left(\alpha-\beta-\frac{D}{2}\right)}{(4\pi)^{D/2}\Gamma(\alpha)\Gamma\left(\frac{D}{2}\right)}(\tilde{m}^2)^{\frac{D}{2}-\alpha+\beta},$$

where $\tilde{m}^2 = m^2 - i\epsilon$. We also recall that

$$\int \frac{d^D p}{(2\pi)^D}(p^2)^\alpha = 0, \tag{A.6}$$

for every α.

We have $\Gamma(x+1) = x\Gamma(x)$, $\Gamma(n+1) = n!$ and

$$\Gamma\left(\frac{n}{2}\right) = \sqrt{\pi}\frac{(n-2)!!}{2^{(n-1)/2}},$$

$$\Gamma(z) = \frac{1}{z} - \gamma_E + \mathcal{O}(z), \tag{A.7}$$

$$\Gamma(z) = \sqrt{\pi}\left[1 + \left(z-\frac{1}{2}\right)\psi^{(0)}(1/2) + \mathcal{O}\left(\left(z-\frac{1}{2}\right)^2\right)\right],$$

where $\gamma_E = 0.5772...$ is the Euler-Mascheroni constant, while $\psi^{(0)}(1/2) = -1.96351...$ and $\psi^{(m)}(z)$ are the polygamma functions.

Popular textbooks on quantum field theory and renormalization

— G. 't Hooft and M. Veltman, *Diagrammar*, CERN report CERN-73-09

— M. Veltman, *Diagrammatica. The path to Feynman diagrams*, Cambridge University Press 1984

— C. Itzykson and J.B. Zuber, *Quantum field theory*, McGraw-Hill Inc., 1980

— M.E. Peskin and D.V. Schroeder, *An introduction to quantum field theory*, Westview Press, 1995

— S. Weinberg, *The quantum theory of fields*, vols. I and II, Cambridge University Press, 1995

— L.S. Brown, *Quantum field theory*, Cambridge University Press, 1992

— J.C. Collins, *Renormalization*, Cambridge University Press, 1984

Papers

Batalin-Vilkovisky formalism

— I.A. Batalin and G.A. Vilkovisky, Gauge algebra and quantization, Phys. Lett. B 102 (1981) 27

— I.A. Batalin and G.A. Vilkovisky, Quantization of gauge theories with linearly dependent generators, Phys. Rev. D 28 (1983) 2567, Erratum-ibid. D 30 (1984) 508

— I.A. Batalin and G.A. Vilkovisky, Existence theorem for gauge algebra, J. Math. Phys. 26 (1985) 172

Ward identities

— J.C. Ward, An identity in quantum electrodynamics, Phys. Rev. 78, (1950) 182

— Y. Takahashi, On the generalized Ward identity, Nuovo Cimento, 6 (1957) 371

— A.A. Slavnov, Ward identities in gauge theories, Theor. Math. Phys. 10 (1972) 99

— J.C. Taylor, Ward identities and charge renormalization of Yang-Mills field, Nucl. Phys. B33 (1971) 436

Index

www.ingramcontent.com/pod-product-compliance
Lightning Source LLC
Chambersburg PA
CBHW031835170526
45157CB00001B/307